KU-541-163

THE NEW CASE FOR GOLD

ALSO BY JIM RICKARDS

The Death of Money: The Coming Collapse of the International Monetary System

Currency Wars: The Making of the Next Global Crisis

THE NEW CASE FOR GOLD

James Rickards

PORTFOLIO
PENGUIN

PORTFOLIO PENGUIN

UK | USA | Canada | Ireland | Australia
India | New Zealand |South Africa

Portfolio Penguin is part of the Penguin Random House group of companies
whose addresses can be found at global.penguinrandomhouse.com.

First published in the United States of America by Portfolio/Penguin, a member
of Penguin Group (USA) Inc. 2016
Published in Great Britain by Portfolio Penguin 2016
001

Set in 10.54/17.20 pts Sabon LT Std
Printed in Great Britain by Clays Ltd, St Ives plc

A CIP catalogue record for this book is available from the British Library

ISBN: 978–0–241–24835–5

To my mother, Sally Rickards, who taught me
something more valuable than gold—love

The twelve gates were twelve pearls, each of the gates made from a single pearl; and the street of the city was of pure gold, transparent as glass.

REVELATION 21:21

CONTENTS

FOREWORD

It's fitting you're reading Agora Financial's special edition of Jim Rickards' *The New Case for Gold*.

"There are only two ways investors know something about gold," Jim told me at the outset of our partnership. "You either studied economics before 1973," which Jim did at the Johns Hopkins School of Advanced International Studies . . . or you've read an Agora newsletter."

At the time he said that, he couldn't have possibly been aware of the special role he'd play in everyday readers' lives, or that he'd continue a long tradition of the financial newsletter industry.

You may be aware that on April 5, 1933, President Franklin Delano Roosevelt signed Executive Order 6102 "forbidding the Hoarding of gold coin, gold bullion and gold certificates within the continental United States."

What you might not be aware of, however, is that that ban

lasted for 42 years. It ended in 1975, while Jim was finishing school, largely because of one blindly determined newsletter man, James Blanchard. It was Blanchard who set up the National Committee to Legalize Gold and worked hard to get the ban lifted by publishing *The Gold Newsletter.*

With the help of other stubborn newsletter editors, like C.V. Myers in Canada—who covertly sent precious metals to daring individual investors in the U.S.—men like Blanchard brought about the legalization of personal gold ownership in the U.S. and the popularization of gold as the ultimate form of savings. This was a huge step toward financial freedom for individual investors, but one that barely made a dent in the mainstream case *against* gold.

This is the torch that Jim Rickards has picked up and carried forward with his monthly *Strategic Intelligence* newsletter and now with this book. A year and a half after Jim joined us in our publishing endeavor, over 100,000 paid subscribers receive *Jim Rickards' Strategic Intelligence* in more than five languages across eight countries.

By owning gold, these investors are prepared in a world barreling toward monetary reset. All thanks to Jim's persuasive explanations. Even more readers will join their ranks now, thanks to this book.

If I can presume to guess what the late newsletter giants of old would think of Jim's defense of gold, they'd be proud. And astounded, too—as they never would have conceived of the 21st-century reasons Jim's outlined for owning physical gold, like the growing threat of cyberfinancial warfare.

This is the reason Jim's case for gold is "new." It isn't that the 21st century has introduced *different* reasons to own gold than before, only *more* reasons. I stress that because the "old" case for gold was never disproven—it just was obfuscated.

Around the same time Jim entered Johns Hopkins as an undergrad, the eight central banks that made up what was called the London Gold Pool abandoned the definition of $35 as the value of one troy ounce of gold. Experts said that this allowed the price of gold to "float." All that really happened, however, was that paper currencies were allowed to run wild. Gold is a constant. As currencies changed relative to one another, gold priced in those currencies moved too.

This caused the *perception* of gold as a stable monetary unit to shift to it as a volatile investment that produced no yield.

Yet here's something most people don't know. If you had owned gold starting on Aug. 15, 1971, when Nixon closed the gold window, you would've done just as well as if you had bought and held stocks and reinvested dividends.

In fact, you would've done better, because you would've enjoyed the extra free time that came with owning gold. Unlike stocks, gold requires no time for active management or care.

When you toil over investing in stocks to "beat the market," on the other hand, you don't just pay with your money—you pay with your life, too. Owning gold, you could've gone fishing, spent time with family or read countless good books, instead of worrying about your financial performance.

The reason is simple. Gold is not naturally money, but money is naturally gold. It is all-weather, tested over thousands of years. It's universally accepted in exchange through good times and bad.

Modern examples of this abound from the collapse of the London Gold Pool in March 1968 through present day . . .

- *Between 1989–1991, the Soviet Union sold off or swapped virtually its entire gold reserves in an effort to sustain its credit*
- *Gold was used in Vietnam when the U.S. shut off dollar balances to be cleared through New York*
- *China was forced to liquidate some of its gold in the aftermath of the massacre in Tiananmen Square, in 1989, when the world community cut off credit lines*
- *Iraq and Libya used gold to get around sanctions . . . as did Iran and Turkey.*

The fact that central banks put down gold while holding massive hoards of it tells you all you need to know.

In *The New Case for Gold*, Jim uses the mainstream gold memes in a sort of financial jiujitsu to prove that gold is money and worthwhile owning. This will become clearer as the world careens toward the third major crash of this century.

One parting thought before you start savoring Jim's book, since he and I share tastes in TV shows . . .

Rod Serling wrote an episode of *The Twilight Zone* called "The Rip Van Winkle Caper." The episode begins in 1961. A group of thieves rob $1 million in gold bricks from Fort Knox. They take the money to a secret cave in the desert. Once there, they use sleep chambers to hibernate until 2061, thinking that when they wake up 100 years later, no one will remember the heist and they'll be rich.

When each wakes up, greed slowly takes hold, as one thief lets his conspirators die of thirst or exhaustion in the desert. The sole survivor he thinks he'll be rich. But he dooms himself because no one can help him drag the heavy sacks of gold bars through the barren wasteland.

As the thief's on the verge of death, a couple in a hovercar drives by...

They stop to help the thief, who's crawling on the ground about to expire. He offers them his gold, if only he can have a drink of water.

The episode ends with the modern man coming back to his wife in the car . . .

"Can you imagine that?" the man says holding a gold bar. "He offered this to me as if it was really worth something."

"Sure, about a hundred years or so ago," the wife replies, "before they found a way of manufacturing it..."

The faith that men can master economic forces greater than themselves is the fatal conceit of our time.

Half a century later, that delusion is tearing at the seams.

Gold has provided unsophisticated governments with more stability than paper has provided sophisticated governments. My best guess is that by 2061, even the intelligentsia won't be able to refute it. Jim Rickards and *The New Case for Gold* will have played an instrumental part.

Far from gold being manufactured and worthless, Jim says, "You won't be able to get gold at any price" when the next crisis hits.

The solid case for why you should own gold sooner, rather than later, is in your hands. Ignore the proof at your own peril.

Regards,
Peter Coyne
Publisher, *Jim Rickards' Strategic Intelligence*

INTRODUCTION

"Gold is a barbarous relic." How many times have you heard this saying before? If, like me, you write and speak publicly about gold, you have heard it a thousand times. It is part of a well-rehearsed litany of those with complete confidence in paper money, and no confidence in gold. As soon as anyone utters a kind word for gold, out comes this robo-response of the good and great fans of fiat.

The antigold reflex is intergenerational. Among the older generation are PhDs who came of age in the wake of famous gold bashers such as Milton Friedman. This generation includes Paul Krugman, Barry Eichengreen, Nouriel Roubini, Martin Feldstein, and others who cover the ideological spectrum from left to right. Friedman's other theoretical contributions are mostly obsolete (it turns out that floating exchange rates are suboptimal, and money velocity is not stable), yet that has done nothing to tarnish how his acolytes perceive gold.

These antigold giants are now joined by a younger generation educated (or miseducated) to believe that gold has no place in a monetary system. This group includes prominent bloggers and commentators such as Barry Ritholtz, Matt O'Brien, Dagen McDowell, and Joe Weisenthal. The antigold crowd does not care if you wear gold rings or watches, although they look at you with condescension leavened with pity when you reveal that you actually own gold bullion. Still, they show their fangs at the mere mention of a gold standard. That's beyond the pale. They are ready, like a carbine with a hair trigger, to fire off arguments on why a gold standard cannot work, will not work, and has always failed. Epithets like "antediluvian" and "Neanderthal" are thrown in for good measure.

This book, *The New Case for Gold*, makes the argument that gold *is* money, that monetary standards based on gold *are* possible, even desirable, and that in the absence of an official gold standard, individuals should go on a personal gold standard, by buying gold, to preserve wealth.

As I set out to make the case for gold, it's comforting to know I'm not alone. Along with the new generation of gold critics comes a new generation of smart, thoughtful gold advocates. These "young guns of gold" include Ronni Stoeferle and Mark Valek in Vienna; Jordan Elieso and Janie Simpson in Sydney; Jan Skoyles in London; "Koos Jansen" (real name: Jan Nieuwenhuijs) in The Netherlands; and Elaine Diane Taylor in Vancouver. It's quite a network! They are a constant source of encouragement and fresh insights for me.

Yet, before we make the case *for* gold, it would be salutary to demolish the case *against* gold first. The next time a gold basher resorts to a reflex response, you will know how to stop him in his tracks with facts, not clichés.

What is the case against gold? The critics know it by heart. Here's the bill of particulars:

1. *Gold is a "barbarous relic," according to John Maynard Keynes.*
2. *There is not enough gold to support finance and commerce.*
3. *Gold supply does not grow fast enough to support world growth.*
4. *Gold caused the Great Depression.*
5. *Gold has no yield.*
6. *Gold has no intrinsic value.*

Each of these assertions is obsolete, false, or ironically an argument in favor of gold. That does not stop the fiat money ideologues from asserting them. Let's dissect these one at a time.

Gold Is a "Barbarous Relic," According to John Maynard Keynes

This is an easy one to rebut. *Keynes never said it.*

What he did say was more interesting. In his book *Monetary Reform* (1924), Keynes wrote, "In truth, the gold standard is already a barbarous relic." Keynes was not discussing *gold* but rather a *gold standard*, and in the 1924 context he was right. The notoriously flawed gold exchange standard that prevailed in various forms from 1922 to 1939 should never have been adopted, and should have been abandoned long before it died with the outbreak of the Second World War.

Keynes was above all a pragmatist. In July 1914, at the start of the First World War, Keynes was the most persuasive voice for remaining on the classical gold standard that had prevailed since 1870. Most countries entering the war abandoned the gold standard immediately in order to finance the war with the gold they retained. His Majesty's Treasury and the Bank of England wanted to do the same.

Keynes argued, in effect, that gold as money was finite, yet credit was elastic. By remaining on the gold standard, and upholding London's role as the center of global finance, the United Kingdom's credit would be enhanced. London could borrow the money needed to finance the war.

That is exactly what happened. The House of Morgan in New York organized massive loans for the United Kingdom and none for Germany or Austria. This finance was critical to Britain's ability to hold out until the United States entered the war in 1917. Victory came the following year.

In 1925, as chancellor of the exchequer, Winston Churchill contemplated returning the United Kingdom to the gold standard at the prewar parity. Keynes told Churchill this would be a deflationary disaster. Keynes did not support the gold standard. Still, he insisted that if Britain was going to have one, it was crucial to get the price right. Keynes recommended a much higher price for gold. Churchill ignored Keynes's advice. The result was massive deflation and depression in Great Britain, years before depression struck the rest of the world.

In July 1944, near the end of his life, Keynes argued at Bretton Woods for a new form of world money he called the bancor, the theoretical predecessor to today's special drawing right (SDR). The bancor would be backed by a commodity basket including gold. It was not a strict gold standard. Still, it did give gold an important place in the monetary system. Keynes's plan was pushed aside in favor of a dollar-gold standard championed by the United States, which lasted from 1944 to 1971.

In short, Keynes was an advocate for gold early in his career, an astute adviser on gold in mid-career, and an advocate for gold again late in his career. In between, he properly disparaged the operation of a flawed gold exchange standard. Keep Keynes's nuanced view of gold in mind the next time someone throws the phrase "barbarous relic" at you.

There Is Not Enough Gold to Support Finance and Commerce

While this claim is nonsense, we will address it because it is one of the most prevalent fallacies shared by fiat money fans.

The amount of gold in the world is always fixed at a level, subject to an increase through mining. Currently the world has about 170,000 metric tons in total, of which about 35,000 metric tons are official gold held by central banks, finance ministries, and sovereign wealth funds. That gold can support any amount of world finance and commerce under a gold standard *at a price*. The price can be determined by calculating the simple ratio of physical gold to money supply.

Assumptions are needed to do this calculation. Which currencies will be included in the gold standard? Which money supply (M0, M1, et cetera) should be used for this purpose? What gold-to-money ratio is best? These are legitimate policy questions that central banks have answered differently over time.

From 1815 to 1914, the Bank of England conducted a successful gold standard with about 20 percent gold backing for the money supply. From 1913 to 1965 the Federal Reserve was required to have at least 40 percent gold backing for the U.S. money supply. Generally, the more confidence people have in the central bank, the lower the amount of gold backing needed to maintain a stable gold standard.

With these inputs, it is possible to calculate the implied gold price for each set of assumptions. For example, if the United States, the Eurozone, and China were to agree on a gold standard using M1 as the money supply and 40 percent gold backing, the implied price of gold would be about $10,000 per ounce. If the same three entities used M2 as the money supply with 100 percent gold backing, the implied gold price would be about $50,000 per ounce.

A gold standard launched at $1,100 per ounce would either be highly deflationary (if money supply were reduced as needed) or highly unstable (as people rushed to buy cheap gold from the government).

The short answer to this criticism is that *there is always enough gold* for a gold standard, as long as you specify a stable, nondeflationary price.

When critics say "there's not enough gold" what they really mean is that there isn't enough gold at current prices. That is not an objection to a gold standard. It is an objection to a candid confrontation with the real value of paper money relative to physical gold. That confrontation will take place as confidence in paper money is eroded, and a gold standard gains favor as a way to restore confidence in our economic system.

Gold Supply Does Not Grow Fast Enough to Support World Growth

This is another canard that reveals a lack of understanding about how gold standards work.

A critic who advances this argument fails to distinguish between stocks of *official* gold and *total* gold. Official gold is owned by the government and available to support a money supply. Total gold includes official gold plus all the gold held privately as bullion or used in jewelry or for decorative purposes.

If a government wants to increase its official gold supply to support monetary expansion, all it has to do is print money and buy private gold on the open market. New mining output is not a constraint. Official gold supplies could double by acquiring private gold, if that was needed, and it would barely make a dent in the total gold stock in the world. (Official gold is only about 20 percent of the total gold stock, which leaves ample room for governments to acquire gold.)

Printing money to buy gold under a gold standard is just another form of open-market operation. It's no different from printing money to buy bonds, which the Federal Reserve does every day. Of course, it does have market consequences, and discretionary monetary policy can lead to blunders. This is true with or without a gold standard. Ultimately, new mining output does not limit central banks' ability to expand credit under a gold standard.

More to the point, consider the following average annual growth rates from 2009 to 2014:

Global GDP	2.9%
World Population	1.2%
Gold Production	1.6%
Federal Reserve Monetary Base	22.5%

Which growth rate is the outlier?

It is true that global GDP grows faster than new gold production. If that were the only factor (it is not), the world could still grow in real terms at its maximum potential (subject to nonmonetary structural impediments), and nominal prices would have a slight deflationary bias. Mild deflation is a benefit for consumers and savers.

There is no reason why a gold standard cannot be combined with discretionary monetary policy. A combination of gold and central bank money was the norm from 1815 to 1971 except during war. Central banks acted as a lender of last resort and expanded or contracted the money supply as they saw fit even under the gold standard. In fact, gold's main purpose was to signal the proper monetary policy based on bullion inflows and outflows.

When critics say gold production does not support world growth, what they really mean is that gold production does not support *inflationary* world growth. This is true. Inflation transfers wealth from rich to poor, from savers to debtors, and from citizens to government. Inflation is the preferred policy of socialists and progressives who favor income redistribution. The objection to gold based on mining output is not that it stands in the way of growth, but that it stands in the way of theft.

Gold Caused the Great Depression

Actually, the Great Depression was caused by incompetent discretionary monetary policy conducted by the U.S. Federal Reserve from 1927 to 1931, a fact documented by a long line of monetary scholars including Anna Schwartz, Milton Friedman, and more recently, Ben Bernanke.

The Great Depression was then prolonged by experimental policy interventions launched by Herbert Hoover and Franklin Roosevelt. These experiments gave rise to what scholar Charles Kindleberger called "regime uncertainty," which meant that large corporations and wealthy individuals refused to commit capital because they were uncertain about regulatory, tax, and labor policy costs. Capital went to the sidelines and growth languished.

Bernanke's own research shows that at no time during the Great Depression was the money supply ever constrained by the gold supply. The law then allowed the Fed to create money up to 250 percent of the value of gold held by the Fed. The actual money supply never exceeded 100 percent of the value of the gold. This means that the money supply could have more than doubled without gold's acting as a restraint. The problem with money supply growth was not gold; it was the fact that customers did not want to borrow and banks did not want to lend. There was a bank credit and consumer confidence problem, not a problem with gold.

On the international side, scholar Barry Eichengreen has pointed out how countries that devalued their currencies against gold (France in 1925, the United Kingdom in 1931, the United States in 1933, and the United Kingdom and France together in 1936) received immediate economic benefit through increased exports. It is true that the devaluing countries received short-term benefit. Still, the world did not. France's benefit in 1925 came at the United Kingdom's expense. The United Kingdom's benefit in 1931 came at the expense of the United States. The benefit to the United States in 1933 came at the expense of the United Kingdom and France.

Eichengreen's otherwise brilliant scholarship suffers from what Keynes called the "fallacy of composition," meaning that what is good for an individual is not necessarily good for an aggregation of individuals. One person standing on her chair at a crowded rock concert may get a better view, yet if everyone stands on their chairs then almost no one can see the show.

The sequence of devaluations against gold from 1925 to 1936 shows the notorious "beggar thy neighbor" currency war dynamic. The crux of the problem was the United Kingdom's decision to return to gold in 1925 at £4.25 per ounce, the pre-World War I parity. Because Great Britain had doubled its money supply between 1914 and 1925, mostly to finance the war, a return to the old parity meant the money supply had to be cut in half. This policy was highly deflationary. The overvalued pound sterling gave France a trading advantage from 1925 to 1931 when Britain finally broke the old parity. The 1931 devaluation gave Britain an advantage, especially against the United States, until 1933 when the United States broke its old parity also.

Gold did not cause the Great Depression; a politically calculated gold price, and incompetent discretionary monetary policy, did.

For a functional gold standard, gold cannot be undervalued (the United Kingdom in 1925, and the world today). When gold is undervalued, central bank money is overvalued, and the result is deflation. A gold standard can work fine, so long as governments set gold's price on an analytic rather than political basis.

Gold Has No Yield

This statement is true, and it is one of the strongest arguments *in favor* of gold.

Gold has no yield or return because it is not supposed to.

Gold is money, and money has no yield because it has no risk. Money can be a medium of exchange, a store of value, and a unit of account, but true money is *not* a risk asset.

To illustrate this simple yet elusive point, just look at a dollar bill. Is it money? Yes. Does it have a yield? No.

Yield comes from putting the dollar in the bank. But then it's not money anymore; it's a bank deposit. (The Federal Reserve defines bank deposits as part of the "money supply." That's because the Fed is in the business of propping up that particular monetary illusion.)

A bank deposit is not money; it's a bank's unsecured liability. The largest banks in the United States would have collapsed in 2008 if not for government bailouts in the form of expanded deposit insurance, guaranteed money market funds, zero interest rates, quantitative easing, foreign central bank swap lines, and other monetary gymnastics. Bank depositors in Cyprus in 2013, and Greece in 2015, received a painful education in the difference between a bank deposit and money. In both cases, banks were closed, ATMs were shut down, and paper currency was soon in short supply. In Cyprus, some depositors had their deposits forcibly converted to bank stock. In Greece, local credit cards could not be processed and a quasi-barter economy quickly emerged.

You can also receive yield by buying stocks, bonds, real estate, or other nonmoney assets. Still, there is risk in doing so. Although many investors think of stocks, bonds, and real estate as money, they are in fact risk assets, just like a bank deposit.

A gold coin, a dollar bill, and bitcoin are three forms of money. One is metal, one is paper, and one is digital. None of them has a yield. They're not supposed to—they're money.

Gold Has No Intrinsic Value

When a reporter or blogger attacks your support of gold by saying it has "no intrinsic value," you should compliment him on his firm grasp of Marxian economics.

The *intrinsic value theory* is an extension of the *labor theory of value* first advanced by David Ricardo in 1811, and later adopted by Karl Marx in *The Communist Manifesto* (1848) and *Das Kapital* (1867, 1885, 1894), among other writings. The idea is that the value of a good derives from the combination of labor and capital that went into its production. The more labor that was needed to produce a good, the more "valuable" it was.

Marx's central critique of capitalism was that bourgeois capitalists controlled the "means of production" and did not pay labor their fair share of the value added. By this means, capitalists extracted "surplus value" from labor. Marx theorized that eventually this exploitation of labor by capital would lead to extreme income inequality, increasing class consciousness among labor, causing a proletarian revolt and the overthrow of capitalism, which would be replaced with a socialist system. Marx's analysis rested on a foundation of intrinsic value.

The problem with this critique of gold is that economic theories based on intrinsic value have not been credited by economists since 1871. That was the year Carl Menger, at the University of Vienna, introduced the idea of *subjective value*. Menger's insight became the cornerstone of what was later known as the Austrian School of economics.

Subjective value is the value assigned to a good by an individual based on that individual's needs and wants. This value is completely independent of any intrinsic value determined by inputs or factors of production. Any gold miner filing for bankruptcy because the market price of gold did not cover his production costs can attest to the irrelevance of intrinsic value.

Gold has almost no industrial uses. It is useful as money, and not good for much else. (Jewelry is not a separate use for gold. It may be visually attractive and pleasing to the wearer. Still, it is wearable wealth—a claim with which any Indian bride will gladly concur—and therefore a form of money in its role as a store of value.) Under the theory of subjective value, the gold price will vary based on gold's utility to someone in want or need of money.

Any exchange economy that has advanced beyond the Robinson Crusoe stage of simple manufacturing and subsistence farming has a need for money. There are many forms of money including gold, dollars, euros, bitcoin, and at certain times and places, feathers, shells, and beads. The value of each form of money varies with the subjective wants and needs of each individual in the economy. At times, dollars may prove highly useful, gold less so, and the dollar price of gold will fall based on this subjective valuation. At other times, confidence in dollars may wane and the dollar price of gold could rise dramatically.

The "intrinsic value" of gold is an obsolete concept, as Menger showed 145 years ago. Anyone raising it as an objection to gold today is wedded to older Marxian economics.

Of these six best-known objections to gold as money, five are empirically, analytically, or historically incorrect, and one—gold has no yield—is correct, yet it's not a critique, it's a truism, and consistent with the view that gold is money.

This is not to say there are no problems with the use of gold as money. Every monetary standard has its problems. The creation of a new gold standard, for example, would require extensive technical work on issues of parity with other currencies and maintenance of those parities. Such a task would resemble the eight years of research that went into the convergence of multiple European currencies in the euro between the Maastricht Treaty (1992) and the euro's official launch (1999).

Yet the gold critics should offer real objections (if they have any), not the red herrings listed above.

Turning from the case against gold, we now consider the case in favor. Unfortunately, some of the most common arguments in favor of gold are as tired and unsubstantiated as the arguments against it.

For example, some conspiracy fans claim there is no gold stored in Fort Knox. Really? If gold were as valuable as they claim, why would the U.S. government let the gold out of its sight?

In fact, the bulk of the U.S. gold hoard is safe and sound in Fort Knox, Kentucky, and West Point, New York, with much smaller amounts at the Denver Mint and the Federal Reserve Bank of New York. This gold may be leased to others through the Bank for International Settlements in Basel, Switzerland, yet that does not mean U.S. gold is not in U.S. custody. Gold leasing is a paper transaction with no requirement for physical delivery.

Other gold supporters claim the fact that the United States has not audited its gold hoard proves the gold is not there. In fact, it proves the opposite. The U.S. government has a powerful interest in downplaying gold's importance. The government wants its citizens to forget their gold even exists (while keeping more than eight thousand metric tons in deep storage). It's also important to note that audits are reserved for important assets, not trivial ones. By refusing to do an audit, the government maintains the pretense that gold is trivial. An audit pays respect to gold's value—that's the last thing the government wants.

This book is called *The New Case for Gold*, with emphasis on the word *new*. Our goal is not to rehearse the same old arguments, but rather to put the gold discussion in a twenty-first-century context. This includes gold's role in cyberfinancial warfare, gold's importance in economic sanctions on nations such as Iran, and gold's future as a competitor to the world money called special drawing rights (SDRs) issued by the International Monetary Fund.

Now let's leave the gold critics and the more tendentious conspiracy fans behind, and set out to explore gold's importance in the hyperactive, digital world we live in today. That is an intriguing journey.

Chapter 1

GOLD AND THE FED

Unemployment, the precarious life of the worker,
the disappointment of expectation, the sudden loss of savings,
the excessive windfalls to individuals, the speculator,
the profiteer—all proceed, in large measure,
from the instability of the standard of value.

JOHN MAYNARD KEYNES, *MONETARY REFORM* (1924)

Is the Fed broke? The colloquial "Fed" refers to the consolidated *Federal Reserve System,* consisting of twelve separate regional Federal Reserve Banks each owned by private banks in its region. By "broke" we mean *insolvent*—an excess of liabilities over assets leaving a negative net worth. Definitions aside, the question remains: is the Fed broke?

I've had occasion to discuss that question with members of the Fed's Board of Governors, regional Federal Reserve Bank presidents, senior Fed staffers, and presidential candidates, among others. The answers I received were "no," "yes," "maybe," and "it doesn't matter." Each of those answers reveals a troubling aspect of the Federal Reserve. Let's look at those answers in turn, and what the people providing them actually meant.

At a superficial level, the Fed is not insolvent. A glance at the Fed's balance sheet, at the time of this writing, shows total assets of approximately $4.49 trillion, total liabilities of approximately $4.45 trillion, and total capital (assets minus liabilities) of approximately $40 billion. Admittedly, the Fed is highly leveraged (about 114 to 1). Of course, leverage amplifies the impact of gain or loss on the capital account. It would take only a 1 percent loss on the Fed's assets to completely wipe out its capital. In normal stock and bond markets, 1 percent declines happen all the time. The Fed's balance sheet is highly leveraged and hanging by a thread yet not technically insolvent.

This brings us to the concept of "mark-to-market." As the name implies, marking to market means taking each asset and repricing it to the current market price using the best available information. Hedge funds and broker-dealers do this every day, although the results are reported only periodically. Banks also use mark-to-market with part of their balance sheets; some assets are marked to market, some not, depending on whether the assets are held for trading or long-term investments.

The Fed does not use mark-to-market accounting. What if it did? Would that make it insolvent? The answer requires a deeper dive into the details of the Fed's balance sheet.

Short-term instruments such as ninety-day Treasury bills hardly vary in price at all. They are not volatile enough to have a major impact on Fed solvency even if marked to market. This is not true of ten-year notes and thirty-year bonds. Both instruments are highly volatile. In fact, the volatility (what is technically called "duration") increases at lower levels of interest rates. Of course, rates have been near all-time lows for the past six years, so that makes these instruments especially vulnerable to large swings in market value.

The Fed balance sheet lumps "U.S. Treasury securities—Notes and Bonds, nominal" into a single category and shows holdings of approximately $2.3 trillion as of this writing. The Fed then breaks those holdings down by regional reserve bank. Of the $2.3 trillion held by the system, $1.48 trillion are held on the books of the Federal Reserve Bank of New York. That makes sense because the New York Fed runs open market operations for the entire system and is the principal buyer of Treasury debt under the various quantitative easing programs (known as "QE"). In turn, the New York Fed provides a detailed list of all Treasury securities owned under its System Open Market Account (SOMA). Using this detailed information on the securities, a daily price ticker, and some conventional bond math, it is possible to mark this portion of the Fed's balance sheet to market.

The New York Fed data shows the Fed was making massive purchases of highly volatile ten-year notes during the height of QE2 and QE3. For QE2, these purchases were made from

November 2010 to June 2011. For QE3, the purchase period was September 2012 to October 2014.

Using this data alone, the Fed was technically insolvent on a mark-to-market basis at certain times from June to December 2013. In that period, the ten-year note had a yield to maturity of about 3 percent. Most of the Fed's ten-year note purchases were at yields of 1.5 percent to 2.5 percent. That backup in yields from the 1.5 percent level to the 3 percent level produced massive mark-to-market losses on that portion of the Fed's portfolio—more than enough to wipe out the slender $60 billion capital cushion at that time.

In late January 2013, just as yields were starting to spike higher, I had dinner at a friend's home in Vail, Colorado. Joining us at the dinner table was a recently resigned member of the Board of Governors of the Federal Reserve who had been on the board during all of QE1, QE2, and the start of QE3. I'm not one to tiptoe around sensitive subjects, so after some cordial conversation, I turned to the former governor and said, "It looks like the Fed is insolvent." The governor seemed taken aback and said, "No, we're not." I explained, "Well, not technically, but on a mark-to-market basis it looks that way." The governor said, "No one has done that math." I replied, "I've done it, and I think others have also."

I looked the governor in the eye and saw a slight flinch. The reply came: "Well, maybe." Then a pause and, "If we are, it doesn't matter; central banks don't need capital. Many central banks around the world don't have capital." I said, "I take your point, Governor. Central banks don't technically need any capital. Still, that might come as a surprise to the American people. There's good reason to believe Fed solvency could be an issue in the 2016 presidential campaign." At that point, I could see our dinner host was getting antsy, so the conversation moved on to more congenial topics like wine and skiing.

My point was not to get bogged down in technical accounting methods and the theory of central banking. My point was that the entire edifice of the Federal Reserve and the dollar rests on a single point of failure—confidence.

As long as confidence is maintained, the money printing can go on. As soon as confidence is lost, no amount of money printing can save you. My concern is that the Federal Reserve is so dominated by MIT-trained quants and PhDs that the policy makers get lost in the models and lose sight of the temperament of the American people and the trust Americans place in them.

In early 2015, I had a private dinner in midtown Manhattan with another Fed official. This time it was not a governor, but an academic specialist handpicked by Ben Bernanke and Janet Yellen to handle Fed policy communications. He was not a public relations professional or anyone the public was much aware of. He was the ultimate insider with an office across the hall from Bernanke and Yellen at Fed headquarters on Constitution Avenue in Washington, D.C. (a fact Bernanke confirmed to me personally when I spoke to him later).

Again, I raised the topic of Fed insolvency on a mark-to-market basis. At that time, ten-year note yields had fallen back below 2 percent and a lot of the ten-year notes purchased in 2010–2013 had lower volatility because they had only five to seven years remaining to maturity. (A ten-year note with five years to maturity trades like a five-year note in terms of duration and volatility.) It seemed likely that the Fed had recouped its mark-to-market losses and was probably solvent at the time of our discussion. Still, I wanted to pursue the topic because rates could rise again, causing new market losses. I was interested in the subject of confidence.

Now the reaction was less equivocal than what I had encountered in Vail. In fact, my friend was categorical: "We're not insolvent, and never have been. It's all on the balance sheet. Have a look." In making this statement, he referred specifically to the period of higher interest rates in mid-2013. That did not faze him. "We've never been insolvent." Case closed.

Having done the mark-to-market math on the bond portfolio, I was intrigued. What was I missing? Did the Fed have some hidden asset that offset the bond losses? It was clear that my friend was leading me in that direction yet did not want to say so explicitly.

I went back to the Fed balance sheet and found what I was looking for right away. In fact, it was the first line on the balance sheet, called the "gold certificate account." As of this writing, that account was listed on the balance sheet at $11 billion. That line entry is the historic cost, the Fed's usual accounting convention. What if that were marked to market just like the bonds?

Understanding the gold certificate account involves a trip back in time to 1913, with a stop in 1934. When the Federal Reserve was established in 1913, its private owners, the banks in each district, were required to transfer their gold to one of the regional reserve banks. This was the first step in transferring physical gold into fewer and fewer hands, a topic we will return to.

In 1934, the U.S. government effectively seized all the gold from the Federal Reserve and moved it into the hands of the U.S. Treasury. Fort Knox was built in 1937 partly to hold the Fed bank gold, and partly to hold other gold confiscated from the American people in 1933.

Pursuant to the Gold Reserve Act of 1934, gold certificates were issued by the U.S. Treasury to the Federal Reserve System, both to plug a hole in the balance sheet and to overcome constitutional objections based on the Fifth Amendment provision, ". . . nor shall private property be taken for public use, without just compensation." The Treasury took the Fed's gold yet gave "just compensation" in the form of gold certificates.

Those gold certificates were last marked to market in 1971 at a price of $42.2222 per ounce. Using that price and the information on the Fed's balance sheet, this translates into approximately 261.4 million ounces of gold, or just over eight thousand tons. At a market price of $1,200 per ounce, that gold would be worth approximately $315 billion. Because the gold is held on the Fed's balance sheet at only about $11 billion, this mark-to-market gain gives the Fed a *hidden asset* of more than $300 billion.

Adding $300 billion to the Fed's capital account reduces Fed leverage from 114 to 1 to a much more respectable 13 to 1, the capital ratio for most well-capitalized banks. This hidden asset is more than enough to absorb the mark-to-market losses on the bond portfolio when they arise.

It is also interesting to note that the amount of gold held by the Treasury, about eight thousand tons, is roughly equivalent to the amount of gold claimed by the Fed on its balance sheet; also about eight thousand tons. The U.S. gold supply dropped from about twenty thousand tons in 1950 to about eight thousand tons in 1980. The drop of twelve thousand tons came in two phases. About eleven thousand tons were lost to dollar redemptions by foreign trading partners from 1950 to 1971. Then another thousand tons were dumped on the market for price suppression from 1971 to 1980. Suddenly the price suppression scheme using United States' physical gold was abandoned, and the United States has engaged in almost no official sales since 1980.

Is this because the Treasury is afraid to hold less gold than it theoretically owes to the Fed? Is eight thousand tons a floor on the U.S. gold hoard because that's how much the Fed claims on its balance sheet? If so, that relationship is highly significant because it means the United States *cannot* dump any more physical gold on the market. It can only encourage others, such as the United Kingdom, to dump their gold or play the paper gold game through leasing operations. The Treasury is out of the game as a source of supply.

Technically, the gold certificates do not give the Fed the right to demand physical gold from the Treasury. They do carry an implied moral obligation that in the event of a collapse of confidence in the Fed's printed money, the Treasury will use its gold to support the Federal Reserve. Another name for an implied obligation to support the Fed with gold is a gold standard.

My friend, the insider, was correct. The Fed was briefly insolvent in 2013 on a mark-to-market basis, if one were to look at its securities portfolio only. Yet it was *never* insolvent when taking into account the Fed's *hidden gold assets*.

The confidence of the entire global financial system rests on the U.S. dollar. Confidence in the dollar rests on the solvency of the Fed's balance sheet. And that solvency rests on a thin sliver of . . . gold. This is not a condition anyone at the Fed wants to acknowledge or discuss publicly. Even a passing reference to the importance of gold to the Fed's solvency could start a debate on gold-to-money ratios and related topics the Fed left behind in the 1970s. Nevertheless, gold still matters in the international monetary system. This is why central banks and governments keep gold in their vaults despite their public disparagement of its role.

Chapter 2

GOLD IS MONEY

People are fascinated by gold not because it is shiny, but because it is money. Understanding this fact is the starting place for understanding gold.

There are many kinds of money in the world, of course. At times, different forms of money have competed for a role as the leading global reserve currency. Today, the dollar, the euro, and bitcoin are all forms of money. So is gold.

What Is Money?

A classic definition of money has three parts: medium of exchange, store of value, and unit of account. If all three of those criteria are met, you have money of a sort. If you ask economists, "What is money?" they reflexively assume that only fiat currencies printed by central banks qualify and lapse into technical discussions about narrow or expanded versions of money supply called M3, M2, M1, or M0, which are all different. Each "M" is narrower than the one before. M0 is the narrowest, consisting of bank reserves and currency. M0 is also called "base money" because it is the narrowest definition of money economists know. I call gold "M-Subzero" because even if economists don't recognize it, it is the real base money behind the paper money supply.

Why Gold?

Gold bashers are quick to disparage gold as a "shiny metal" or "pile of rocks," as if to say it has no particular attraction as a form of money. Even sophisticated economists such as former Federal Reserve chairman Ben Bernanke have described gold's continued storage in U.S. vaults as a "tradition," with no suggestion that there was anything more useful to say about it.

In fact, the use of gold as money is not only ancient but eminently practical. Recently Justin Rowlatt of the BBC World Service conducted an interview with Andrea Sella, professor of chemistry at University College London, in which Professor Sella provided an in-depth review of the periodic table of the elements to explain why gold, among all the atomic structures in the known universe, is uniquely and ideally suited to be money in the physical world.

We all recall the periodic table of the elements from high school chemistry class. It looks like a matrix of squares, one square per element, about eighteen squares wide and nine squares high, with an irregular shape around the edges; hydrogen (H) and helium (He) stand above their peers. Each square contains the name of an element, and its one- or two-letter symbol, along with useful information such as the atomic weight, atomic mass, and boiling point. A total of 118 elements are represented this way, from hydrogen (atomic number 1) to ununoctium (temporary name for atomic number 118). The important point for our purposes is that there is nothing physical in the known universe that is not made of one of these elements or a molecular combination. If you're looking for money, you'll find it here.

Professor Sella deftly leads us through a tour of the table. He shows that most of the matter in the universe is completely unsuitable for money. He then zeros in on that handful of elements that are suitable and singles out the one that is nearly perfect for the purpose—gold.

Sella quickly dismisses ten elements on the right-hand side of the table, including helium (He), argon (Ar), and neon (Ne). The reason is obvious—they're all gases at room temperature and would literally float away. They're no good as money at all.

In addition to the gases, Sella rejects elements such as mercury (Hg) and bromine (Br) because they're liquid at room temperature, and are as impractical as the gases. Other elements such as arsenic (As) are rejected because they're poisonous.

Next he turns to the left-hand side of the table, which includes twelve alkaline elements such as magnesium (Mg), calcium (Ca), and sodium (Na). These are no good as money either because they dissolve or explode on contact with water. Saving money for a rainy day is a good idea, but not if the money dissolves as soon as it rains.

The next elements to be discarded are those such as uranium (U), plutonium (Pu), and thorium (Th), for the simple reason that they're radioactive. No one wants to carry around a form of money that might cause cancer. Also included in this group are thirty radioactive elements made only in laboratories that decompose moments after they are created, such as einsteinium (Es).

Most of the other elements are also unsuitable as money based on particular properties. Iron (Fe), copper (Cu), and lead (Pb) don't make the final cut because they rust or corrode. It's bad enough when central banks debase your money. No one wants money that debases itself.

Rowlatt and Sella continue on their trip through the periodic table. Aluminum (Al) is too flimsy to use as coins. Titanium (Ti) was too hard to smelt with the primitive equipment available to ancient civilizations.

Once the process of elimination is complete, there are only eight candidates for use as money. These are the so-called noble metals, situated about in the center of the table, consisting of iridium, osmium, ruthenium, platinum, palladium, rhodium, silver, and gold. All of these are rare. Still, only silver and gold are available in sufficient quantities to comprise a practical money supply. The rest are extremely rare, too rare to be money, and difficult to extract because of very high melting points.

Rowlatt completes his *tour d'horizon* this way:

> *This leaves just two elements—silver and gold. Both are scarce but not impossibly rare. Both also have a relatively low melting point and are therefore easy to turn into coins, ingots and jewelry. Silver tarnishes—it reacts with minute amounts of sulphur in the air. That's why we place particular value on gold.**

It turns out gold (Au) has one final attraction—it's golden. All of the other metals are silvery in color, except copper, which turns green when exposed to air. Beauty is not a prerequisite for money. Still, it's a nice attribute for gold considering it passes every other test with flying colors.

Our ancestors did not use gold just because it was shiny or beautiful as modern critics suggest. Gold is the only element that has *all* the requisite physical characteristics—scarcity, malleability, inertness, durability, and uniformity—to serve

* Justin Rowlatt, "Why Do We Value Gold?," *BBC World Service Magazine*, December 8, 2013, www.bbc.com/news/magazine-25255957.

as a reliable and practical physical store of value. Wiser societies than ours knew what they were doing.

Of course, this list of virtues does not mean that gold *has* to be money. Today's money exists mostly in digital form. Electrons that store the digits don't rust either. Then again, they're not the least bit scarce.

Just because money is "digital" doesn't mean it's not part of the physical world. There is no escape from the periodic table of the elements. Digital money exists as charged subatomic particles stored on silicon (Si) chips. Those charges can be hacked and erased. Gold atoms (atomic number 79) are stable and cannot be erased by Chinese and Russian cyberbrigades. Even in the cyber age, gold still stands out as money nonpareil.

Gold Is Not an Investment

Gold is not an investment, because it has no risk and no return. Warren Buffett's well-known criticism of gold is that it has no return and therefore no chance of compounding his wealth. He's right. Gold has no yield; it's not supposed to, because it has no risk. If you buy an ounce of gold and keep it for ten years, you end up with an ounce of gold—no more, no less. Of course, the "dollar price" of a gold ounce may have changed radically in ten years. That's not a gold problem; it's a dollar problem.

To get a return on an investment, you have to take risk. With gold, where is the risk? There is no maturity risk, because it's just gold. It won't "mature" into gold five years from now; it is gold today, and always will be. Gold has no issuer risk, because nobody issues it. If you own it, you own it. It's not anyone else's liability. There's no commodity risk. With commodities there are other risks to consider. When you buy corn, you have to worry: Does it have bugs in it? Is it good corn or bad corn? It's the same with oil; there are seventy-five grades of oil around the world. But pure gold is an element, atomic number 79. It's always just gold.

Gold Is Not a Commodity

Gold has almost no industrial uses. It is not a commodity, because it's not an important input to any production process, save a few. Look at any other commodity: copper is used for wires and pipes, and silver actually has many industrial uses in addition to being a precious metal. Other mining commodities are used as inputs in manufacturing and production; gold is not. Gold has uses in electronics for coatings, connections, and the like, but they are still limited—nothing material.

We know that gold is traded on commodity exchanges and reported in the commodity section of your favorite Web site. Breathless reporters describe gold price action from the commodities trading pits. Still, that does not make gold a commodity. That's important for investors to understand, because there are many developments that affect commodities that do not affect gold in the same way.

Consider the situation in the Great Depression. The most daunting economic problem was deflation. Commodity prices and industrial production dropped precipitously. Yet from 1929 to 1933 the U.S. dollar price of gold was not deflating; it remained fixed at $20.67 per ounce. Gold was performing a monetary role, not a commodity role.

In a matter of months, beginning in April 1933, the U.S. government forced the gold price higher, from $20.67 an ounce to $35.00 an ounce. The government raised the gold price to cause inflation; they were desperate to break out of deflation, and gold led the way by government fiat. Stock and commodity prices soon followed. Gold did not act at all like a commodity; it acted like money. Today, governments again fear deflation, and seek inflation to help reduce the real burden of sovereign debt. Gold may again be enlisted to catalyze the inflation that central banks have thus far failed to produce.

Another example of gold's noncommodity behavior can be seen in the correlation of gold to the Continuous Commodity Index in 2014. That index has sixteen components, including gold as well as iron ore, copper, aluminum, and agricultural commodities. Gold exhibited a high degree of correlation to the index from January through November of that year, which is to be expected. But in November 2014, the index plunged and gold rose sharply in dollar terms. This divergence coincided with plunging energy and base metals prices (accounting for the index move), and growing Russian and Chinese demand for gold (accounting for the gold move). Gold had suddenly stopped trading like a commodity and began trading like money. Such behavior is the shape of things to come.

Gold Is Not Paper

Wall Street sponsors, U.S. banks, and other members of the London Bullion Market Association (LBMA) have created enormous volumes of "gold products" that are not gold. These are paper contracts.

These products include exchange-traded funds—ETFs—the most prominent of which trades under the ticker symbol GLD. The phrase "ticker symbol" is a giveaway that the product is not gold. An ETF is a share of stock. There *is* some gold out there somewhere in the structure, but you don't own it—you own a share. Even the share is not physical; it's digital and easily hacked or erased.

The legal structure behind GLD is a trust, and the trustee has some physical gold in a vault. This arrangement is generally true of gold ETFs. The GLD vaults are in London. There is a set of authorized participants who make a market in the GLD trust shares. These are the large LBMA members such as Goldman Sachs, JPMorgan Chase, and others.

Much of the authorized participant activity consists of arbitrage between the physical gold market and the market for GLD shares. If there is selling pressure on GLD shares, the authorized participant can buy the shares as a market maker and sell physical gold short. Then the dealer can deliver the shares to the trustee, receive physical gold in exchange, and cover the short physical position, thereby pocketing a profit measured by the difference between the share price and the physical price. Such arbitrage is similar to the "gold points" arbitrage that existed between New York and London in the years prior to 1914, except that it is no longer necessary to ship physical gold across the North Atlantic to make a profit. Nowadays, the gold just sits in an LBMA vault or the GLD vault depending on the flow of the arbitrage.

Investors in GLD take other risks besides not having physical gold and being targets of digital hacking. For example, officials could close the New York Stock Exchange, leaving investors unable to trade the shares. Those who say the exchange will never be closed should recall that it happened due to a software glitch on July 8, 2015, during Hurricane Sandy in 2012, and after 9/11. Famously the New York Stock Exchange was also closed over four months at the outbreak of World War I. Power outages or electronic problems could also close the exchange again at any time. With an ETF you're locked into that digital system.

The London Bullion Market Association also sells gold through paper contracts that act like unregulated futures. The gold underlying these contracts is described as "unallocated," which means the owner has no claim to any particular physical gold. The seller does have some physical gold, but still not enough to satisfy the potential claims of all the unallocated gold buyers. Banks can sell $10 or more of these contracts for every $1 of physical gold they hold. They hope that all the holders don't show up at once and ask for the gold, because if they do, they're not going to get it.

Gold owners under these contracts have to give notice to the bank if they want to convert from unallocated gold to gold that is specifically allocated and held in custody. The notice period gives the bank time to locate some physical gold to cover the contract.

If too many customers claimed physical gold at once, the bank could terminate the contract and simply provide the counterparty with cash at the closing price as of the termination date. The customer would get a check at that closing price, but they're not going to get the actual gold. That's the best case. In a worst case, the bank could fail and the gold investor would get nothing.

So these paper contracts might offer price exposure to gold markets, yet that's a far cry from actual gold. If there's a demand shock or a buying panic for gold and the price of gold is skyrocketing, that's exactly when these paper contracts are going to fail, because there will not be enough physical gold to satisfy all the claims. Only physical gold in nonbank custody is real gold.

Gold Is Not Digital

Gold is a physical, not digital, currency. As such, gold provides insurance against the risks to which digital currencies are exposed.

For the most part, the dollar is a digital currency. We may have a few paper dollars in our pockets, but not many relative to our needs. If I go to the grocery store, I may pull out a twenty-dollar bill, but I'm more likely to pull out my debit card.

When you get your paycheck, it's probably a direct deposit to your account from your employer. When you pay your bills, you likely use online banking. When you go shopping, you probably use a credit or debit card. The amount of cash you're using is tiny relative to the volume of your economic transactions.

The largest securities market in the world, the United States Treasury market, has not issued a physical paper certificate since the early 1980s. There might be a few old paper certificates floating around in someone's attic, but the Treasury bond market today is completely digital, as is the payment system.

The cashless, digital society is already here. Some observers are concerned about what they call "the war on cash." Don't worry—the war on cash is over and the government won.

As a practical matter, honest citizens cannot get access to large quantities of cash without being suspected of drug dealing, terrorism, or tax evasion. Along with that suspicion comes government surveillance. Citizens without gold have no choice but to go along with the digitization of wealth.

Digital wealth is subject to power outages, infrastructure and exchange collapses, hackers, and online theft. What good is even a billion-dollar portfolio if it can be wiped out overnight?

What if the government shuts down the banks and reprograms the ATMs to limit you to $300 a day for gas and groceries? The fact that you might have $100,000 in the bank is irrelevant. Government regulators will say $300 a day is more than enough for gas and groceries until further notice.

This is the exact scenario that played out in the Eurozone in Cyprus in 2013 and Greece in 2015. Savers should have physical gold as insurance against a bank-freeze scenario.

The History of Monetary Collapse and the End of the Gold Standard

Gold is money. Still, its status as money has been disparaged continually by governments and economists, especially in the period since the international monetary system collapsed and America ended the convertibility of dollars into gold in 1971. The monetary collapse in 1971 should have come as no surprise. The international monetary system actually collapsed three times in the twentieth century—1914, 1939, and 1971—and came perilously close to collapse in 1998 and 2008.

Because today's international monetary system is largely based on the U.S. dollar, a new collapse will be triggered by a collapse of confidence in the dollar and its role as a store of value. It may be surprising. Still, such collapses do happen every thirty years or so. Based on the monetary history of the past century, we're probably at the end of the useful life of the current international monetary system and fast approaching a new one.

Prior monetary collapses have not meant the end of the world. People did not go into caves and start eating canned goods. Monetary collapse meant that the major financial and trading powers of the time sat down around a table and rewrote what they called the "rules of the game," which is a shorthand expression for the operation of the international monetary system.

For example, after the 1914 collapse, there was a monetary conference in 1922 in Genoa, Italy, where the major powers rewrote the rules of the game and attempted to reintroduce the gold standard. After the 1939 collapse, there was a larger, well-known international monetary conference in Bretton Woods, New Hampshire, in 1944 that rewrote the rules of the game around a dollar-gold standard. Then after the collapse in 1971, when President Nixon suspended the convertibility of dollars for gold, there was a series of conferences, the most famous being the Smithsonian Agreement in December 1971. Numerous subsequent negotiations took place all the way up through the Plaza Accord in 1985 and the Louvre Accord in 1987, which rewrote the rules of the game once again.

The period from 1971 to 1980 was one of temporary chaos as the United States muddled through and moved toward floating exchange rates. It was a dreadful period of economic performance. The United States had three recessions between 1973 and 1981. The dollar price of gold went from $35 an ounce to $800 an ounce. Inflation took off. The dollar's value was cut by more than half.

The dollar was rescued by Paul Volcker and Ronald Reagan beginning in 1981. This is when the world moved to a new "dollar standard," also known as the King Dollar period.

In effect, the United States told the world that even in the absence of a gold standard, the dollar would be a reliable store of value. This meant ending dollar inflation and making the United States an attractive destination for dollar investments. Volcker's monetary policy and Reagan's tax and regulatory policies accomplished these goals. U.S. trading partners were essentially told they could anchor to the dollar. The sound dollar standard was successful from 1981 to 2010, a period characterized by solid growth until 2007, and with long economic expansions in the 1980s and 1990s.

So, from 1870 to 1971, the international monetary system used variations on a gold standard with interruptions for wars. For thirty years, from 1980 to 2010, the world did not have a gold standard. We had a dollar standard instead. Now we have no standard and no anchors whatsoever in the international monetary system. It should come as no surprise that since 2007 we have been living with confusion, volatility, and suboptimal performance in the markets and the economy.

When the next collapse comes, there will be another such meeting as those held in Genoa in 1922 and Bretton Woods in 1944. Investors today need to look ahead and ask, "What will the new rules of the game be?" Based on the answers, they can figure out how they should construct their portfolios today to protect their net worth when the inevitable turmoil happens.

Gold Never Went Away

It is generally believed that President Nixon closed the gold window on August 15, 1971, and the United States has been off the gold standard ever since. And two generations of students have since been rigorously conditioned by policy makers and professors to believe that gold has no role in the international monetary system.

The truth is, gold has never gone away. The power elites stopped talking about it and publicly ignored it, yet they held on to it. If gold is so worthless, why does the United States have more than eight thousand tons? Why do Germany and the IMF keep approximately three thousand tons each? Why is China acquiring thousands of tons through stealth and Russia acquiring more than one hundred tons a year? Why is there such a scramble for gold if it has no role in the system?

It's highly convenient for central bankers to convince people that money is unconnected to gold because that empowers them to print all the money they want. Everyone from Ben Bernanke to Alan Greenspan and others have disparaged gold, saying it plays no part in the system. Along with the power to control money comes the power to control behavior and politics. Still, gold is the foundation, the real underpinning, of the international monetary system.

Gold and the International Monetary System

Gold is making its comeback in the world monetary system. When you look at what's actually going on in the world as opposed to the happy talk you hear on television, it is clear that the world is already on a shadow gold standard and is moving back to a more formal gold standard—treating gold as money. We're seeing signs of that already; it's not just a possibility for the distant future. The evidence that gold is moving back toward the system center is clear, and it's happening for a number of reasons.

The International Monetary Fund (IMF) is the third largest gold holder in the world. (Number one is the United States, number two is Germany, and number three is the IMF. It is likely that China is actually the second largest gold holder, although its true holdings are not publicly disclosed and are difficult to confirm.)

The IMF plays a pivotal role in the global monetary system, with far more power and influence than one might assume based on its technocratic and bureaucratic demeanor. The IMF likes to posture as a benign friend to small, emerging countries. In reality, it's more like a large, rapacious corporation that makes a donation to charity every now and then just to show how generous it is.

The IMF was created at the Bretton Woods Conference in 1944. It took a few years to get it up and running in the late 1940s and early 1950s. It started as a swing lender for wealthy countries experiencing short-term balance-of-payments deficits.

Consider a country running a balance-of-payments deficit year after year. One of the ways to fix the deficit would be to cheapen its currency to make exports more competitive. But cheapening the currency wasn't allowed under the Bretton Woods fixed exchange rates. Instead the IMF would provide a loan to tide you over while you made structural reforms to your economy. Such reforms would attempt to lower unit labor costs, improve productivity, or improve the investment climate—whatever was necessary to get the current account back toward a surplus. Once the capital account was in surplus, the swing loan from the IMF could be repaid.

In extreme cases, the IMF allowed devaluation, but only after all other monetary and structural solutions had been exhausted.

This swing lending system broke down in the late 1960s and early 1970s when the United Kingdom steeply devalued sterling against the dollar, and the United States suspended gold convertibility. The fixed exchange rate system died soon afterward. Since then, we've had floating exchange rates.

After the 1980s, the IMF wandered in the wilderness for almost twenty years with an uncertain mission. In the 1980s and early 1990s it acted as a lender to emerging markets, because its original mission of stabilizing exchange rates under Bretton Woods was gone.

The IMF's reputation suffered badly in the 1997–98 Asia financial crisis. There was blood in the streets—and not metaphorically. People were killed in riots in Jakarta, Indonesia, and Seoul, South Korea. Many people to this day, most famously Nobel Prize winner Joe Stiglitz, attributed this financial crisis to bad advice from the IMF.

By 2000, the IMF was like a whale that had washed up on the shore and couldn't get back to its mission in the sea. No one quite knew what they were doing and how they should be doing it. By 2006, there were public calls to abolish the IMF.

Then a funny thing happened on the way to the IMF's demise. We had a global financial crisis in 2008, and suddenly, the IMF was back in the game. It became the de facto secretariat of the G20 club of the most powerful developed and emerging markets countries. The G20 acts as a kind of board of directors with the IMF as a staff and agency to implement the will of the board.

The IMF has its own governing board, yet interestingly, if you look at that membership by country, there's a lot of overlap with the G20 membership. The G20 member countries and the twenty-four member countries of the IMF executive committee are, by and large, the same. The G20 was a group of heads of state that didn't have a staff while the IMF has a ready-made staff. Since 2009, the G20 summits have worked hand in glove with IMF technical capabilities, staff, and analysts. New lending facilities have become highly politicized as seen in Ukraine and Greece.

The fact is the IMF has always been a rich countries' club. The IMF voting mechanism requires an 85 percent vote to make any significant changes, such as a change in its articles (the IMF's governing document), or to approve a major lending initiative. The United States has over 16 percent of the votes, which means that if all the other members combined voted against the United States, it still wouldn't be enough to overcome U.S.

objections. None of this is by coincidence, of course; the United States has always been the largest voice in the IMF, and the IMF headquarters building is in Washington, D.C.

One of the big governance issues in the international finance world now involves changing these voting arrangements. If you approach the issue in terms of how much your GDP is as a percentage of world GDP, and compare that with your IMF vote, rich countries are overweighted, and emerging markets are underweighted. China is a good example. China accounts for about 14 percent of global GDP, yet its vote in the IMF until recently was less than 5 percent. Legislation to give China a larger voice was passed by the U.S. Congress in 2015. Recognition of China's rightful place in the IMF hierarchy is a quid pro quo by the United States in exchange for China's good behavior in the battle over currency manipulation against the U.S. dollar.

Now the IMF is back to its original mission of lending to rich countries, mostly bailing out Europe, where the overwhelming majority of its money is directed. The bulk of the IMF's money is not going to poorer countries like Botswana or Mali or Jamaica. It's going to Poland, Greece, Portugal, Ireland, and, for political reasons, Ukraine.

This new lending spree requires new sources of funding for the IMF itself. If you're going to lend money, where do you get the money to lend? Banks can take deposits, pledge assets to central banks, or create money out of thin air. The

IMF doesn't have a teller window where you can make a deposit. Still, it does borrow the money. It issues notes. Interestingly, these notes are not denominated in dollars. They're denominated in special drawing rights (SDRs), worth about $1.38 each as of this writing, although the SDR value fluctuates with the market.

What is an SDR? Well, it's world money. But not the kind you carry around in your pocket. You can't go to an ATM and withdraw a bundle of SDRs. Still, SDRs are money, and they play an increasingly important role in global finance as the dollar's power declines. There's actually a trading desk inside the IMF that can swap SDRs for other hard currencies. Here's a simple example of how that works. In 2009 the IMF issued 182.7 billion SDRs, equivalent to about $255 billion at current exchange rates. The way it issues them is in accordance with a quota, which is simply the word it uses for a country's share. If I have a 5 percent quota at the IMF and the IMF is issuing 100 billion SDRs, then I'm going to get 5 billion SDRs or 5 percent of the total issued. Many IMF members had quotas yet didn't need the SDRs, and wanted other hard currencies instead.

Hungary is a good example. Going back to the early 2000s, Hungarian banks offered customers mortgages in two currencies. They could take a loan in local currency, which is the forint, or they could take the loan in Swiss francs provided by European banks in Vienna or Zurich that could fund the loans. Swiss franc mortgages were about 2 percent and forint mortgages were about 9 percent, so most of the borrowers took Swiss franc mortgages assuming the exchange rate would remain fixed. But it didn't. The forint collapsed, and suddenly the mortgage debt relative to the borrowers' local income increased dramatically. Defaults skyrocketed.

If you're Hungary and the IMF gives you SDRs, your reaction is that you really need Swiss francs so your central bank can help the local banks repay the interbank loans. Now you call the IMF trading desk and say, "Offer me dollars for my SDRs." The IMF desk will call China and say, "Do you have a bid for SDRs?" China says, "Yes, we do." China will send dollars to the IMF and get SDRs in return while Hungary will get the dollars, sell them to buy Swiss francs, and then use the francs to help their banks. That's the way to turn your SDRs into another hard currency if you need it.

The IMF doesn't issue SDRs except in liquidity crises. The next time there is a global liquidity crisis, it will be bigger than the capacity of the Federal Reserve and other central banks to contain. The Fed has used up its balance sheet—used up its dry powder, if you will—dealing with the last crisis. It has not been able to unwind the balance sheet, and it's unlikely it will for a decade. The same is true of the other central banks. They have no further capacity to print money without destroying confidence. They might have the legal capacity to print more, yet they are at the limit of what they can credibly do.

In such a new liquidity crisis the world will turn to the IMF and be reliquefied by the issuance of SDRs. That process may work without impinging on confidence because so few understand it. This massive SDR issuance will be highly inflationary in dollar terms. Still, politicians in Washington will simply point the finger at the IMF as an unaccountable agency.

One effect of a massive SDR issuance will be to hinder capital formation by destroying the real value of dollar-denominated assets. The only shelter in the storm will be hard assets, including gold. Astute individual investors are positioning their portfolios that way today, and so are major powers such as Russia and China.

What if people lose confidence in the IMF and the SDR solution? Who bails out the IMF? Right now there isn't anyone. Turning to the IMF is not kicking the can down the road; it's more like kicking the can upstairs, from private debt to sovereign debt to multilateral debt issued by the IMF. The IMF is the penthouse; you can't kick the can any higher. And the IMF's source of strength is its three thousand tons of gold, and gold held by IMF members in the United States and Europe.

This is why I keep returning to the subject of gold, and why I calculate gold-to-money ratios and gold-to-GDP ratios, and develop dollar price projections of $10,000 per ounce or higher. If confidence in national paper money is lost and you try to bail out the system with a different kind of fiat money, specifically the SDR, what good does that do? If it works at all, it will only be for two reasons. One, almost no one understands it, and two, we won't have SDRs in our pockets. SDRs will be used by, for, and between countries—not by individuals. SDRs won't be transparent. They will exist and be highly inflationary if printed in sufficient quantities. Still, no one will actually see them because they are the most technical and abstract form of money ever created.

If SDRs work, it will be in part because so few people understand them. Still, if people do understand, they are likely to lose confidence. In that scenario, the only recourse is gold.

Shadow Gold Standard

Countries around the world are acquiring gold at an accelerated rate in order to diversify their reserve positions. This trend, combined with the huge reserves held by the United States, the Eurozone, and the IMF, amounts to a shadow gold standard.

The best way to evaluate the shadow gold standard among various countries is to use the ratio of gold to the gross domestic product (GDP). This gold-to-GDP ratio can easily be calculated using official figures and compared across countries to see where real gold power resides.

The big winners—the real center of gold power in the

world—are the nineteen nations that make up the Eurozone and issue the euro. Their gold as a percentage of GDP is over 4 percent. The United States' ratio is about 1.7 percent. Interestingly, Russia's ratio is about 2.7 percent. Russia has more than one eighth the amount of gold the United States has, yet its economy is only one eighth the size of the U.S. economy, so the ratio is higher. Russia is one of those nations that are acquiring more gold, and it seems set on matching the Eurozone. Japan, Canada, and the United Kingdom are major economies, yet their gold ratios are anemic, all less than 1 percent.

The most interesting case is China. The official gold reserves of China are reported as of July 2015 at 1,658 tons. Yet we know from various reliable sources including mining production and import statistics that their actual gold stock is closer to 4,000 tons. I've spoken to refineries and secure logistics firms—people who actually handle physical gold—in addition to official sources, and included their information in my estimates. On the whole, there is enough credible information available to support this estimate at a minimum. It is also entirely possible that China has considerably more than 4,000 tons of gold.

China, like Russia, is acquiring gold so that it will have a comparable ratio to the United States and Europe. The gold-to-GDP ratio will be critical when the monetary system collapses because it will form the basis for any monetary reset and the new "rules of the game."

In any monetary reset, countries will come together, as I have described, and sit around the table. One can think of that meeting as a poker game. When you sit down at the poker table, you want a big pile of chips. Gold functions like a pile of poker chips in this context. It doesn't mean that the world automatically goes to a gold standard. It does mean that one's voice at the table is going to be a function of the size of one's gold hoard.

There are only about 35,000 tons of official gold in the world. The phrase "official gold" means gold owned by central banks, finance ministries, and sovereign wealth funds. This does not include gold jewelry and gold held in private hoards.

China's acquisition of more than three thousand tons of gold in the past seven years represents almost 10 percent of all the official gold in the world; a huge shift in gold reserves in China's favor. This buying program explains China's nontransparency. The gold market is liquid, but thinly traded. If China's intentions and actions were fully disclosed, the gold price would likely be much higher. This is always true when a huge buyer shows up in a thin market. China wants to keep the gold price as low as possible until it completes its acquisition program.

China is trying to acquire enough gold so that when the international monetary collapse comes and the world has to recut the deal, China will have a prime seat at the table. Countries like Canada, Australia, and the United Kingdom with small gold-to-GDP ratios will be seated away from the table, along the walls. These small gold powers will essentially be spectators in the global monetary reset and will have to content themselves with whatever system the United States, Europe, Russia, and China devise. In this scenario, Germany will speak for Europe, so the new system will be based on a U.S.-German-Russian-Chinese monetary condominium administered by the IMF. These major gold powers are already preparing for such an outcome. This is what I mean when I refer to the shadow gold standard.

Conclusion

Gold is money. Despite disparagement by policy makers and economists, it will remain as a store of wealth par excellence, and continue to play an integral part in the world's monetary system. In part we can thank the French, who took a stand at the IMF in 1975 and insisted on a role for gold in official reserves even when it was no longer a monetary reference point at the time.

Academic economists don't seem to care about gold. It is mostly ignored, and never studied in a monetary context. Still, gold has never completely gone away. It still matters behind the scenes. Gold is still poised in the reserves of the international monetary system and will be even more important in the years to come.

Understanding gold provides us with a frame of reference for understanding the future of the international monetary system. In the chapters ahead we'll look at how smart investors are investing in physical gold to protect themselves from the complex economic forces and instability we face in the twenty-first century.

Chapter 3

GOLD IS INSURANCE

As discussed in the last chapter, gold is not an investment, it's not a commodity, it's not a paper contract, and it's not digital. Gold is simple, an element, atomic number 79; it is the opposite of complex. It is robust in the face of international monetary collapse and financial market complexity. Owning gold is insurance against the current economic climate and unstable monetary system.

Gold is the anticomplex asset, and therefore one asset that an investor should own in a complex world. Let's take a look at what I mean by complexity, and consider the ways gold can insure us against complex systemic risk.

Complexity Theory and Systems Analysis

When analyzing the state of the world economy and the potential for collapse, I use complexity models. Complexity is a branch of physics that explores the impact of recursive functions in densely connected networks. It is the science of how nodes are interconnected and how they interact. The interaction leads to changed behavior, also called adaptive behavior, which can produce completely unexpected outcomes. The Federal Reserve, however, uses stochastic equilibrium models, which are not a good representation of how the real world works.

Such technical terms are a bit daunting. Still, these concepts are not all that difficult. So what is an equilibrium model?

A good example I think all investors can relate to is an airplane. An airplane is made of tons of aluminum, steel, and other heavy physical components, and yet it manages to fly at high altitudes. How does it do that? The answer is that an airplane is shaped and engineered in a certain way. The wing is flat on the bottom and curved on top so that more air goes under the wing than on top of the wing; the curvature on top blocks it. The shape provides lift.

How does the airplane get air moving under the wing? It has engines that give it thrust. With thrust and lift, it's up in the air. But now it needs to turn, because the air traffic controller says it needs to go one place instead of another. How does it do that? It uses a rudder. When the plane needs to descend, there are flaps that change the shape of the bottom of the wing. And so on.

Now picture the Fed chair as a pilot in the cockpit with her hands on the controls. She can use flaps to change the shape of the wing, she can use a throttle to give it a little more or less thrust, and she can use a rudder to turn the plane to port or starboard as needed. Let's say there's a little bit of turbulence. The pilot says we're going to go up a little bit higher to get above the turbulence and give the passengers a smooth flight. If the plane is losing altitude, the pilot gives it a little more thrust or lift to make it go up.

The Fed chair is sitting in the boardroom of the Federal Reserve in Washington thinking of the economy as a plane that's not going quite fast enough or not flying quite high enough, so a little more thrust in the form of printing money, maybe a little rudder in terms of forward guidance, and maybe a little lift in terms of quantitative easing will help the plane achieve its objective. That's an equilibrium model.

There's only one problem with this model. The economy is not an equilibrium system. The economy is a complex system. What is a complex system? Imagine that the airplane suddenly turns into a butterfly; that's an example of complexity.

Complexity produces the unexpected, or what's technically called an "emergent property." An emergent property is a development you don't see coming. So here, the Fed is trying to fly the airplane using all the monetary policy tools, but with the complexity risk that the plane performs in completely unexpected ways.

Take the case of the banking system. Banks have never repaired the damage from the 2008 crisis, nor have they fixed the issues that led to the collapse in the first place. You'll hear many commentators and regulators say that bank balance sheets are stronger and bank capital ratios are higher. That's true. Still, they're not strong enough relative to the risks, and the system is still unstable.

The five largest banks in the United States today are larger than they were in 2008. They have a higher percentage of total bank assets, and their derivatives books are significantly larger. The banks that were "too big to fail" in 2008 are bigger and more dangerous today.

When you have a concentration of assets in a small number of banks that all do business with one another, mostly in derivatives, there is a high degree of density. This means that if one small problem arises anywhere in the system, that perturbation will spread through the system rapidly. That is called contagion or, in the jargon of the IMF, "spillover." Whether you call such a financial disturbance contagion or spillover, it has the same domino effect on the banking system.

What is even more daunting about complex systems is that the most cataclysmic outcomes can occur from minute—actually impossible to perceive or measure—changes in initial conditions. It does not take large causes to produce large consequences. Quite trivial events, such as the unexpected failure of a small, unknown broker in a distant part of the world, can cause systemic collapse depending on linkages at the time of the failure.

Consider this metaphor: A mountain has a steeply pitched section near the top. It has been snowing for weeks, and the snow has piled up. There's apparent avalanche danger. Experts can see the windswept snow leaning in an unstable way. The snowpack will clearly collapse at some point.

The snowpack can persist for a while. Perhaps the more daring skiers want to ski underneath it because the view is nice or because they're daredevils.

One day a snowflake falls, hits the mountainside, and disturbs a few other snowflakes. This disturbance starts a small chute that turns into a bigger slide. It gathers momentum, brings more snow with it, and creates force. Suddenly the whole mountainside is destabilized and comes crashing down, killing skiers in its path and burying a village below.

When we go back and do a postmortem, whom do we blame? Do we blame the snowflake or do we blame the unstable snow conditions? Of course we need to blame the unstable snow conditions, because even though a single snowflake started the avalanche, it was going to happen anyway. If it wasn't that one particular snowflake, it could have been the one before or the one after. It's the instability of the mountainside that gave rise to the avalanche and the destruction in its path. The mountain snowpack was a complex system waiting to collapse.

Here's another example. Let's say you're in a theater with one hundred people listening to a presentation. Suddenly two people get up and run out of the room. What do you do? What does everyone else do? Odds are you do nothing. You may think their behavior is strange or rude. Perhaps the people who ran out got an urgent text or were late for an appointment. Regardless, you're going to sit there and watch the rest of the presentation.

Now let's say that instead of just two people, suddenly sixty people got up and ran out of the theater. What would you do? What would the others do? I daresay you'd be right behind them, because you would assume those people knew something you didn't. Perhaps the place was on fire, maybe there was a bomb threat, but you don't want to be the last to know. You flee the theater.

That's an example of adaptive behavior based on a variable called the "critical threshold." The critical threshold is the point at which the behavior of others affects your behavior. In the example above, your critical threshold (T) for fleeing the theater is greater than two and less than sixty, which can be represented mathematically as: $2 < T < 60$.

Everyone in the theater may have a different critical threshold, and those thresholds may change frequently depending on external conditions or your mood or any other factor. The theatergoers may remain calm and seated if just a few people run out. Still, the entire audience may suddenly erupt in panic when more than a few start to run. It's difficult to know just what the tipping point for full-scale panicked behavior is.

To get a sense of the complexity in capital markets, imagine applying this simple example not to a hundred people in a theater, but to a hundred million investors around the world transacting in capital markets, in foreign exchange, in commodities, in stocks and bonds, and in derivatives every single day.

If you're a stock market investor and you see the market going down, you might say it represents a good buying opportunity. It goes down more, and you say, I see a lot of value here. Now it goes down more and you're losing a fortune. At what point do you throw in the towel? At what point do you panic? At what point do you say, "You know what, I'm getting out of here. I'm selling my stocks!" Your selling may drive the market down further and cause still more investors to sell. The selling just feeds on itself. That's an example of how massive changes of outcome can be catalyzed by small changes in initial conditions.

It doesn't take a lot. It just takes a snowflake or a few people changing their minds to affect others. Momentum builds, and eventually everyone is running out of the theater panicking, or capital markets are collapsing. Most people don't see it coming. A good working knowledge of complexity theory and complex systems dynamics at least helps you to understand the dangers.

The best approach is not to focus on individual snowflakes but to study the systemic instability. By getting a good grasp of complexity, you can anticipate systemic collapse even without seeing the snowflake.

As long as the Federal Reserve clings to equilibrium models and fails to make use of complexity theory, it will continually miss bubbles and underestimate systemic risk as it has done repeatedly over the past thirty years. High intelligence and a PhD in economics are no substitutes for good modeling. When you apply the wrong model you will get the wrong result every time.

International Networks

Complexity theory is the most important new tool in economics today, not only for understanding U.S. monetary policy, but for understanding global capital markets as well. Because capital markets are complex systems, not equilibrium systems, every central bank macro model in the world is obsolete. It should come as no surprise that we keep having crises and meltdowns.

This is not a recent phenomenon. Consider 1987, when the stock market fell 22 percent in a single day, equivalent to about 4,000 Dow points from today's level. If the stock market fell 400 Dow points today, it would dominate news coverage and public discourse. Imagine if it fell 4,000 points in a single day. In effect, that's what happened in 1987.

In 1997 we had the Asian meltdown, in 1998 the Long-Term Capital Management meltdown, in 2000 the tech stock meltdown, in 2007 the mortgage meltdown, and in 2008 outright financial panic. Why do these crises keep happening? The reason is that the Fed is sitting there in the cockpit trying to fly the airplane, but the economy is not an airplane—it's much more complex.

If you're trying to implement policy in a complex system using an equilibrium model, you'll get it wrong every time. So it's important for the rest of us to understand complexity and see how it applies to capital markets. That's a more fruitful endeavor than following Fed policy debates.

We can be sure that there are connections, interrelations, and spillover effects because that's the nature of networks. If you apply graph theory and network science to the way financial nodes are actually laid out, that conclusion is inescapable. The problem is that these connections can be difficult to see in real life. I'll give you a concrete example.

I was in Tokyo in September 2007 right after the U.S. housing markets started to crash. The panic peaked in 2008 with Lehman Brothers and AIG. Still, the crisis really started in the summer of 2007. As Tokyo's stock market was falling, my Japanese colleagues could not initially see the linkages. They understood there was a mortgage problem in the United States but did not see what that had to do with Japanese markets.

I explained to them that when you're in financial distress, you sell what you can, not what you want. In this particular case, what was happening was that hedge funds and other leveraged investors in the United States were getting margin calls on their bad mortgages. They would have loved to sell their mortgages, yet there was no market for mortgages or any other asset-backed securities at the time. So they started selling Japanese stocks—not because they hated Japanese stocks, but because Japanese stocks were liquid and could easily be sold to raise cash to meet the margin calls on other positions. Even though the two markets do not usually correlate with each other, distress in U.S. mortgage markets caused a significant simultaneous decline in the Japanese stock market. This is what my former colleague and Nobel Prize winner Myron Scholes called "conditional correlation." It's a correlation of two markets that does not usually exist yet suddenly comes into existence upon the happening of a specified condition. Conditional correlation is a perfect example of what a physicist would recognize as an emergent property in a complex system.

Here's another example. The United States is the world leader in satellite technology, not only for communications and entertainment but also for military and intelligence applications. Boeing is a major firm in that field. Boeing manufactures the satellites in the United States yet outsources the satellite launches to Russia. Since 2014, animosity between Russia and the United States has grown, primarily due to the situation in Ukraine. If we let that tension go too far and start cutting off trade and other business relationships with each other, suddenly you can't launch new missiles in Russia and intelligence eyes in the sky go dark. The dwindling of space-based U.S. intelligence capabilities thus has hidden linkages to rising geopolitical tensions between the United States and Russia involving Ukraine. This linkage may not have been obvious when escalating tensions began, yet it "emerged" from the complex dynamics.

One analytic problem is that many people don't understand what "complexity" means when used in a technical way. Many throw the word "complex" around as jargon or use it interchangeably with "complication."

In fact, complication and complexity are two different conditions when the terms are used in a technical sense. For example, if you take the back off a Swiss watch, what do you see? There are gears, wheels, springs, jewels, and other components. That's a complicated system to be sure. Still, an expert watchmaker can open the watch, remove a gear, and clean or replace it in order to fix the watch. The watchmaker then closes the case and the watch is as good as new.

Now imagine you take the back off the same watch and instead of gears you find a metallic liquid soup. How do you change a gear now? That's an imaginary example of complexity in which the watch movement has gone through a phase transition from a solid to a liquid. Now the watchmaker's craft fails him. Old models don't work.

It's like a pot full of water on a stove. You turn up the heat, and for a long time the pot is still full of water. Suddenly the water turns to steam. The pot contains the same H_20 molecules, yet the molecules have gone through a phase transition. The water molecules now exist in a different state. The water has gone from a liquid state to a gaseous state.

If you have ever watched a pot of water boil, you know that before the water turns to steam, the surface becomes bubbly and turbulent. If one was to anthropomorphize a molecule in a pot of water coming to a boil, that molecule doesn't "know" if it wants to be water or steam. It thinks "water-steam, water-steam," et cetera—it can't decide which to be. Suddenly the turbulence erupts and the water turns to steam. Still, if you turn the heat down, the surface reverts to water. That turbulent surface is where each water molecule passes from one state (water) to another state (steam).

That is a good metaphor for where the world is today. We're out of an old state that existed prior to 2007 yet haven't arrived at the new state. We're on that turbulent, bubbly surface, and investors are quite confused.

Complexity and Policy

The good news is there's a lot that can be done at the policy level to reduce the risk of capital markets' complexity. The bad news is that policy makers are not taking any constructive steps in that direction. Complex systems collapse because they are unsustainable beyond a certain size or scale. Either the energy inputs are too great to sustain the system or the interactions are too numerous to remain stable, or both. In either case, the remedy is to descale the system to a sustainable level before a collapse occurs.

What does the ski patrol at Aspen Mountain do when they see an avalanche danger? They go out early in the morning before the first skiers arrive, they climb a ridgeline, and they set off dynamite charges. They're actually blowing up the snow and causing it to go down harmlessly before it collapses spontaneously and kills skiers. What does the United States Forest Service do when they see massive forest fire danger? They actually start a controlled fire to burn off the dry wood so that a lightning bolt or campfire doesn't set off an even larger conflagration that does a lot more damage.

The amount of dry wood in a forest or the amount of snow on a mountainside are examples of scaling metrics in complex systems. In capital markets, we have scaling metrics as well. These include measures such as the gross size of derivatives, asset concentration in the banking system, and total assets of the largest banks. These are the financial equivalents of unstable snowpacks and dry forests. Just as forest rangers and ski patrols descale the systems they manage, so regulators should regularly descale the banking system.

We should break up the big banks into smaller units, making them like utilities where they serve a useful function and get paid a fair amount for their functions, but no more than that. Even if we do break up large banks, that doesn't mean they can't fail. It just means that when they fail, it won't matter. The point is not to eliminate failure. The point is to eliminate catastrophic collapse that arises from failure. We should also ban most derivatives and bring back the Glass-Steagall Act, which kept banks out of highly leveraged and risky securities businesses.

Opponents of breaking up big banks argue that size creates efficiencies that reduce the costs of banking services for customers. Yet those efficiencies, what are called first-order benefits, are small compared with the second-order costs of catastrophic collapse.

In other words, bank lobbyists are good at touting the benefits of a big bank in terms of typical economies of scale and global competitiveness, but they totally ignore the second-order costs that are borne by society as a whole. The long-term benefits of not having a collapse will outweigh the short-term costs of descaling the system. That's not a calculus that policy makers are really capable of doing, because they don't understand the complex system dynamics at play.

I don't see any signs that regulators and bankers really comprehend complexity theory. Still, they do seem to realize another systemic collapse is coming. The United States is not getting the growth it needs to pay the debt. Derivatives are piling up, the banks control Washington, and the financial system is failing. Gold is the only sensible insurance in this state of affairs.

Financialization of the Economy

The past thirty years have witnessed the extreme financialization of the economy. This refers to the tendency to generate wealth from financial transactions rather than from manufacturing, construction, agriculture, and other forms of production. Traditionally finance facilitated trade, production, and commerce. It supported other activities, but was not an end in itself.

Finance was a little bit like grease on the gears—a necessary ingredient but not the engine itself. But in the last thirty years, finance has metastasized: it has become like a cancer. It acts as a parasite on productive activity.

At the time of the crisis in 2008, the financial sector of the U.S. economy represented about 17 percent of stock market capitalization and 17 percent of GDP. That's an enormous percentage for a facilitation activity. Why should the banking sector be 17 percent of GDP? It should be perhaps 5 percent, which is closer to its historic share. Now finance has become an end in itself, driven by greed and the bankers' ability to devise arcane ways to extract wealth from this complex society. The difficulty is that the means bankers use to extract wealth add to complexity without adding value. Extreme financialization almost destroyed the global economy in 2008.

As a result of mining, gold stock increases at a fairly steady rate. In the past there were occasional large discoveries, although there have not been any in more than a hundred years. It was in the period from about 1845 to 1898 when there were some large gold discoveries. Since then, annual output has been slow and steady, at about 1.6 percent a year.

Interestingly, gold stock has been increasing at about the

same rate as the global population, resulting in honest money. It's almost as if gold scarcity were heaven-sent for this purpose.

Still, if you have honest money, you can't financialize the economy, because finance couldn't grow any faster than production plus innovation. What the finance industry needs is either leverage or credit instruments: derivatives, swaps, futures, options, and various kinds of notes and commercial paper. You need all of what I call "pseudo money"—a phrase I used in my first book, *Currency Wars*—in order to keep the game going.

Finance does not create wealth; it extracts wealth from other sectors of the economy using inside information and government subsidies to do so. It's a parasitic or so-called rentier activity. Finance needs to be contained before it triggers the next crash. This involves breaking up big banks, banning most derivatives, and constraining the money supply.

The Role of the Federal Reserve

Given its importance, the Federal Reserve is one of the least understood major institutions in U.S. society. The Federal Reserve is a complicated, multitiered system. Most attention is paid to the Board of Governors in Washington. The Board of Governors has seven members, yet there have been vacancies lately, so the Fed board has been operating with as few as three or sometimes four members.

The next tier of the system consists of the twelve regional Federal Reserve banks located in the major economic centers around the country. These include the Federal Reserve Bank of New York, the Federal Reserve Bank of Boston, and the Federal Reserve banks of Philadelphia, San Francisco, and Dallas, and other cities. These regional reserve banks are not owned by the United States government and are not principally agencies of the United States government; they are privately owned by the banks in each region. For example, Citibank and JPMorgan Chase are both in the New York region, and therefore own stock in the Federal Reserve Bank of New York.

When one discusses private ownership, many act like it's a deep, dark conspiracy, yet it has been this way since the Federal Reserve System was created in 1913. The private ownership is well known and not a secret.

The Federal Reserve Bank is privately owned at the regional reserve bank level. Still, control of the entire system resides in the Board of Governors, appointed by the president of the United States and confirmed by the United States Senate. So the system is an unusual hybrid of private bank ownership with government oversight and control.

The Fed's Tools to Execute Policies

The Federal Reserve can directly control what it calls the policy rate or the federal funds rate through open-market operations. These open-market operations consist of purchases and sales of Treasury securities from a network of banks known as "primary dealers." When the Fed buys notes from a primary dealer it creates the money to pay for it out of thin air. When the Fed sells notes to a primary dealer, the dealer pays the Fed and the money disappears. It's that simple.

The Fed has done this for decades. It's really one of its main missions. Open-market operations are conducted by a trading desk at the Federal Reserve Bank in New York, which gives the New York reserve bank a unique role in the overall system.

The New York Fed can sell short-term notes to tighten policy if it wants rates to go up. It can buy these notes, create money, and loosen policy if it wants rates to go down. That's the customary role of open-market operations and is usually conducted using short-term Treasury securities, or the "short end" of the yield curve.

The question for the Fed in recent years has been, how do you control longer-term interest rates? When the fed funds rate is zero and you can't lower it any more, how do you influence medium-term and long-term rates?

There are two approaches to this problem. The first is

simply to buy medium- and long-term Treasury notes. This was done under the title "quantitative easing" or QE, and was an experiment in monetary policy devised by former Fed chairman Ben Bernanke. The idea is that if long-term rates are lower, investors will seek higher yields elsewhere by bidding up stock and real estate prices. These higher prices in stocks and real estate would create a "wealth effect" that made investors feel richer. Based on the wealth effect, investors would spend more money and stimulate the economy. This theory is mostly nonsense. Bernanke's experiment will be viewed in the future as a massive failure. Yet this is how the Fed has behaved since 2008.

The other way to lower medium-to-long-term rates is with the use of "forward guidance." Forward guidance consists of telling markets about future short-term rates. Investors already know that the short-term rate today and tomorrow is near zero. The idea is to identify what the short-term rate will be next year and perhaps the year after that. This is what the Fed was doing when they invented phrases like "extended period" and "patient" to describe interest rate policy.

How does such forward guidance affect interest rates today?

When dealers decide what to pay for a ten-year note today, they don't think of it as a monolithic ten-year interest rate. They think of the note as the present value of a strip of ten one-year forward rates. In effect, they aggregate expectations about a one-year rate in two years, a one-year rate in three years, and so on. So when the Fed gives fresh forward guidance saying short-term rates are going to be low not just today and tomorrow, but a year or two years from now, that has a direct impact on the ten-year-note rate today because dealers change the way they calculate that figure.

Forward guidance, in theory, does have an impact on medium-to-long-term rates, and direct purchases obviously have an impact on those interest rates. The combination of the two is designed to lower rates, which, in theory, causes asset values to go up as described. The process is manipulative, attenuated, and convoluted, yet it is how the Fed operates. Those are the channels. They're using forward guidance and long-term asset purchases to get people to spend more money.

The Fed says they're going to give you clear guidance on the future path of short-term rates, but how do you know they're telling the truth? How do you know they won't change their minds? How credible is the forward guidance? The problem here is that the Fed thinks it's reducing uncertainty about the future path of interest rates, yet all it is doing is replacing one form of uncertainty with another.

Forward guidance is only credible if you actually believe it. Still, considering the fact that the Fed has had fifteen different policies since 2008, it is difficult to know what to believe anymore. It cut rates to zero in numerous stages from 2007 to 2008, and then it did QE1, QE2, QE3, and Operation Twist.

It created numeric targets for unemployment and inflation in December 2012 and then abandoned them when the targets proved worthless. It talked about nominal GDP targeting, and extended the forward guidance over the course of five years in 2011, 2012, 2013, 2014, and 2015.

This is not a disciplined experiment. The Fed is simply making it up as it goes along. If it has fifteen policies in seven years, that's clear evidence of improvisation. Why should investors believe the Fed? One of the reasons forward guidance and Fed policy generally have failed is because the Fed now has no credibility.

Interest Rates: Nominal vs. Real

Real interest rates in dollar debt markets are one of the most powerful influences on the dollar price of gold. Yet the interest rates you hear and read about every day are not *real* rates, they're *nominal* rates. The difference between the real rate and the nominal rate is inflation or deflation. The concept is simple, yet it is widely ignored or misunderstood.

A nominal rate is just the rate that an instrument actually pays you when you buy it. If you buy a new ten-year Treasury note with a 2 percent coupon, then the nominal rate will be 2 percent. After that note is issued, nominal rates may change based on market factors and new note issuance. In that case, the ten-year note you purchased will change in price. If interest rates go up, the price will drop, and if rates go down, the price will rally. This price change then amounts to a premium or discount when you sell the note. The combination of the original coupon plus or minus the effect of the premium or discount gives a "yield to maturity," which is also expressed in nominal terms. Whether you focus on a coupon or a yield to maturity, you are looking at nominal yields.

The real interest rate is just the nominal rate minus inflation. So, if the nominal rate is 5 percent and inflation is 2 percent, then the real interest rate is 3 percent. (5 – 2 = 3). The math is a bit less intuitive when deflation appears. Deflation can be thought of as "negative inflation." When you subtract a negative, you have to add the absolute value to get the result. So if the nominal rate is 5 percent and deflation is 2 percent, then the real interest rate is 7 percent (5 – (–2) = 7)! It gets even less intuitive when nominal rates are negative (as they are in Switzerland and the Eurozone) and deflation appears. Now the math is negative minus negative. For example, if nominal interest rates are negative 1 percent and deflation is 2 percent, then the real interest rate is *positive* 1 percent (–1 – (–2) = 1). The point here is not to conduct a math class, but to illustrate the difference between nominal and real rates and show how the difference may be nonintuitive when negative sums are involved.

Why does this matter to gold investors? It matters because

real interest rates are the alternative to gold. If gold has no yield (true), and if you can earn a low-risk real rate of return elsewhere (sometimes true), then the real rate of return represents the opportunity cost of owning gold. There are other costs to owning gold such as storage, shipping, insurance, and commissions. Still, the real interest rate tends to dominate the others. You might want to own gold even when real rates are high based on expectations, but a high real rate will certainly affect the decision of most investors.

Therefore, the relationship between interest rates and gold is straightforward. High real rates are bad for the dollar price of gold. Low or negative real rates are good for the dollar price of gold. Where are we now? Curiously, both gold investors and the Federal Reserve are on the same side of the trade. Both want negative real rates (albeit for different reasons). The problem for gold investors and the Federal Reserve is that real rates have been persistently high. The Fed and gold investors may want negative real rates, but in the words of Mick Jagger, "You can't always get what you want."

Why does the Federal Reserve want negative real rates? Because it is a powerful inducement to borrow money. Negative real rates are better than a zero interest rate because you can pay back *less* than you borrowed in real terms because the money's not worth as much due to inflation. In a world of negative real rates, almost every project makes sense and the "animal spirits" (in Keynes's famous phrase) of entrepreneurs are aroused. If you borrow money at 2.5 percent and inflation is 3.5 percent, the real rate is *negative* 1 percent; you get to pay the bank back in cheaper dollars. That's the power of negative real rates.

In a world of negative real rates you actually get a *positive* return on gold. If you borrow money to buy gold, and the cost of money is negative, then the zero yield on gold is higher than the negative rate of interest, that is, 0 > -1%. A negative real rate, which prevailed for much of the 1970s, is positive for gold. The 1970s were the period when gold went from $35 per ounce to $800 per ounce in less than ten years.

How does the Fed achieve negative real rates? It has some control over nominal rates using open-market operations as described. Still, the key to negative real rates is inflation. The Fed has used every tool at its disposal to produce inflation including rate cuts, quantitative easing, currency wars, forward guidance, nominal employment targets, and other measures. All have failed. This is because inflation is fundamentally a function of money velocity, or turnover, and that is a psychological and behavioral phenomenon. The Fed has not had any luck changing the deflationary psychology of savers and investors. The Fed is trying to get to negative real rates. The problem is, it's not working.

Inflation and Deflation

To most investors, inflation is intuitive. When price levels take off, expectations change quickly. Views on housing, gold, land, and other inflation hedges start to feed on themselves as investors use leverage to acquire hedges, sending prices even higher.

Deflation is much less intuitive. Deflation has not been a serious economic problem in the United States since the 1930s. Investors and savers are unaccustomed to it and not always aware of its dangers.

Deflation is driven today by demographics, technology, debt, and deleveraging. The real value of debt goes up in deflation, which increases loan losses. Those losses come home to roost at the banks. Because the primary role of the Federal Reserve is to bail out banks, the Fed will do everything possible to stop deflation. Deflation also hurts government tax collections because workers don't get raises, meaning that there is no incremental income to tax. (A worker can still have an improved standard of living in deflation even without a raise because the cost of living goes down. Yet government has not figured out a way to tax a cost-of-living decrease.)

Deflation also feeds on itself. If you expect the price of a good to decline, you may wait to purchase it. That waiting decreases demand in the short run and causes further price declines. The resulting shortfall in aggregate demand can lead to layoffs, bankruptcies, and an economic depression. Deflation *is* a serious threat to government finances, which is precisely why the government will do whatever it can to avoid it.

What's going on in the economy today is best described as a tug-of-war between deflation and inflation.

Deflation is the natural consequence of the debt binge that home buyers and credit card shoppers went on between 2002 and 2007. The resulting debt pyramid came crashing down in the panic of 2008. Deflation is what you would expect in a depression, which the United States has been in since 2008. Such deflation is amplified by deleveraging, asset sales, reducing balance sheets, and other factors.

Inflation is facilitated by central bank policy, mostly money printing, and catalyzed by changes in expectations that lead to increased turnover or velocity of money.

In fact, price indices are showing little change: about 1 percent per year. This is because the forces of inflation and deflation are pushing against each other to some extent, canceling each other out.

In addition to fighting deflation, the Federal Reserve must cause inflation so the United States does not go broke. The national debt is more than $18 trillion as of this writing. The debt does not ever have to be "paid off" in full, yet it does have to be sustainable. The test of sustainable debt is if the economy is growing faster in nominal terms than the debt and interest. Real growth is fine, but real growth is not needed to sustain debt. What is needed is nominal growth, which is real growth plus inflation. Because real growth is so hard to achieve and because the United States continues to incur new debt year after year, the only way out is inflation. Of course, inflation is bad for savers and retirees because their fixed incomes and bank accounts are worth less. Still, inflation is great for debtor countries like the United States because the debt is worth less also. Inflation is the key to making debt affordable.

The economic problem with debt is that it is governed by laws and contracts that look at debt from a *nominal* perspective. If I borrow a dollar from you, I owe you a dollar. In real terms the dollar might be worth $1.50 or maybe 50 cents in terms of purchasing power, depending on whether we have inflation or deflation by the time I pay it back. Still, I owe you a dollar.

The United States owes the world $18 trillion, and in order to pay that off, we are going to need nominal growth of a certain magnitude. Does the Fed prefer real growth? Yes, it does. But will it accept nominal growth with a big inflation component in lieu of that? Yes, it will, if that's the best it can get. In the absence of sufficient real growth, either inflation or outright default is inevitable. The dollar price of gold goes up in either scenario because it is real money.

From 2013 to 2015, the U.S. budget deficit came down significantly, from about $1.4 trillion to about $400 billion. That's a huge drop and good as far as it goes. Yet—and here's the point—the debt-to-GDP ratio continued *going up*, because *there was still a deficit* and nominal growth was not enough to shrink the ratio. The United States is still on the same path as Greece or Japan, although the pace has slowed a bit lately.

In the end, inflation is going to win the tug-of-war because the Fed's tolerance for deflation is so low and the consequences of deflation are so devastating. The Fed must have inflation and will do "whatever it takes," in Mario Draghi's words, to achieve it. Inflation may take time and more rounds of money printing, and forward guidance. Still, it will happen eventually. That inflation will be the catalyst for negative real interest rates and a significant increase in the dollar price of gold.

Insurance Against Inflation and Deflation

It is important for gold investors to understand the difference between real and nominal measures that we use to describe interest rates and Fed policy. Gold investors were dismayed in 2014 and 2015 because gold's dollar price declined, despite financial and geopolitical crises in Greece, Ukraine, and Syria and a stock market crash in China. Gold is supposed to be the "safe haven" asset in times of stress. Why was the price not going up?

A better question would be, why did it not go down more? The dollar price of oil fell over 70 percent from June 2014 to

January 2016, and yet gold's dollar price barely budged over the same period (despite volatility). In fact, the price of gold held up well compared with those of many leading commodities.

Deflation can get out of control, and if it does, it would not be surprising to see gold's dollar price drop further in nominal terms. For example, assume gold is $1,200 per ounce at the start of a year and there is 5 percent deflation that year. Further assume that gold's dollar price at the end of the year is $1,180. In this scenario, the *nominal price of gold fell 1.7 percent* (from $1,200 to $1,180), yet the *real price of gold rose about 3.3 percent,* because the $1,180 year-end dollar price is actually worth $1,240 in purchasing power relative to prices at the beginning of the year.

If the dollar price of gold falls even more sharply, it is likely that other important prices and indices are falling as well. This would be typical in a collapsing or highly deflationary environment. Prices of goods other than gold will decline even more in an extremely deflationary world. If gold is going down in nominal terms, yet other prices are going down more, gold will still preserve wealth when measured in real terms.

Gold may be volatile when measured in nominal dollars. Still, the volatility has more to do with the value of the dollar than with the value of gold. Historically gold has done well in inflation *and* deflation because it represents a real store of value.

Most gold investors have little difficulty understanding why gold does well in an inflationary environment. Yet why does gold also do well in deflationary environments? The reason, as discussed above, is that central banks such as the Fed cannot tolerate deflation. They will do everything possible to create inflation. When all else fails, they can always use gold to create inflation out of thin air by simply fixing the dollar price of gold at a much higher level. Then all other prices will quickly adjust to this new higher gold price. The reason is that the higher dollar price of gold really means that the value of dollars has declined relative to a certain weight of gold. It takes more dollars to buy the same weight. A decline in the value of dollars is the definition of inflation. Government can always fix the dollar price of gold to achieve the inflation it cannot achieve by other means.

This is exactly what the United States did in 1933, and what the United Kingdom did in 1931 when both countries devalued their currencies against gold. In 1933, the U.S. government forced the price of gold from $20.67 per ounce to $35.00 per ounce. It wasn't a case of the market taking gold higher; the market was in the grip of deflation at the time. It was the government taking gold higher in order to cause inflation. The reason the government did that in 1933 was not because it wanted gold to go up; it wanted *everything else* to go up. It wanted to increase the price of cotton, oil, steel, wheat, and other commodities. By cheapening the dollar against gold, it caused inflation in order to end deflation.

In a period of extreme deflation today, the government

could unilaterally take the price of gold to $3,000 or $4,000 an ounce or even higher, not to reward gold investors (although it would), but to cause generalized one-time hyperinflation. In a world of $4,000 gold, all of a sudden oil is $400 a barrel, silver is $100 an ounce, and gas is $7 a gallon at the pump. Price increases of such size would change inflationary expectations and break the back of deflation.

When currency is devalued relative to gold, it works, because gold can't fight back. If the United States tries to devalue the dollar relative to the euro, the Eurozone can fight back by cheapening the euro. But if the United States cheapens the dollar relative to gold (by raising the gold price), that is the end of it. You cannot mystically create more gold to lower the price again. Gold cannot fight back in a currency war.

So we have two paths to higher gold prices—inflation and deflation. It's hard to know which one will prevail, because forces are strong in both directions. The attraction of gold is that it preserves wealth in both states of the world. In inflation, the gold price just goes up as we saw in the 1970s. In deflation, the gold price also goes up, not by itself, but by government dictate as we saw in the 1930s. Gold has a place in every investor's portfolio because it is one of the few asset classes that perform well in both inflation and deflation. That is the best kind of insurance.

Chapter 4

GOLD IS CONSTANT

Price of Gold

When people say the price of gold went "up" or "down," that's only one frame of reference. Instead think of gold as a constant unit of measurement, what economists and mathematicians call a *numéraire* or counting device. On this understanding, it is the currencies that are fluctuating, not the gold. If the dollar price of gold goes from $1,200 an ounce to $1,300 an ounce, most people say gold went "up." My view is that gold didn't go up, the dollar went down. It used to cost me $1,200 to buy an ounce of gold; now it costs me $1,300 to buy an ounce of gold. I get less gold for my money, so the dollar went down.

If you think the dollar is going to get stronger, you might not want so much gold. If you think the dollar is going to get weaker, which I expect over time, then you certainly want gold in a portfolio.

You don't have to correlate gold to another benchmark. Just think of gold as money. The dollar price of gold is simply the inverse of the dollar. If you think of gold as the antidollar, you're on the right track. A strong dollar means a weaker dollar gold price. A weak dollar means a higher dollar gold price. If you're concerned about the dollar—and there's good reason to be concerned—then it makes sense to own gold.

I live in the United States, where I earn and spend dollars. If I buy gold, I'm using dollars to buy the gold. If you live in Japan and earn yen and are going to retire on yen, it's a different equation. Gold can be doing better in yen than it is in dollars if the yen is going down against the dollar. If gold is down 10 percent against the dollar, yet the yen is down 20 percent, then for a yen-based investor, gold is actually up if the yen is your yardstick.

To get a global perspective on gold, you have to analyze not just gold prices in dollars but also cross rates in all currencies. This was the problem in India a couple of years ago. The Indian rupee was crashing against the dollar, and Indian gold sales were slowing down. This was not because Indians had lost interest in gold. If you were paying in rupees at a time when gold was down in dollars, it was actually going up in rupees. That explains why some of the Indian purchases slowed down. It's a complicated world, yet you have to decide what your base currency is and then think about gold in that space, not just in dollar space.

That said, my advice to investors is that when you have gold, you should think about the quantity of gold by weight, not dollar price, and how that fits into your portfolio. Don't get too hung up on the dollar price, because the dollar could collapse quickly and then the dollar price won't matter. What *will* matter is how much physical gold you have.

Still, on a practical level, you do see the dollar price of gold each day. It's quite volatile. That's a reason to keep your gold allocation to about 10 percent of your liquid assets. Gold is an attractive part of a portfolio. Still, it is always prudent to diversify.

The Paper Gold Market vs. the Physical Gold Market

There's another puzzle in this story. Investors assume that prices of goods are driven by the laws of supply and demand. When looking at the global market for physical gold it seems there is a massive increase in demand and no particular increase in supply. Why isn't the gold price responding to this mismatch?

Investors should understand that there's a physical gold market and a paper gold market. The paper gold market consists of a number of contracts: COMEX futures, exchange traded funds (ETFs), gold swaps, gold leasing, forward contracts, and so-called unallocated gold issued by London Bullion Market Association banks. Those derivatives—futures, swaps, ETFs, leasing, forwards, and unallocated gold—form the paper gold market.

The paper market could easily be one hundred times the

size of the physical market. This means that for every hundred people who think they own gold, ninety-nine of them are wrong. Only one of them is going to get physical gold when the panic begins.

That kind of leverage is fine as long as there's a two-way market. As long as price action is not disorderly, as long as people are willing to roll over contracts, and as long as people don't insist on physical delivery, the leveraged paper system works fairly well. The problem is that a lot of those assumptions, even if true initially, can disappear overnight. More investors are demanding physical delivery. Central banks around the world are also demanding physical delivery from custody at the Bank of England or the Federal Reserve Bank of New York. We've seen Venezuela taking back its gold to Caracas, Germany taking back its gold to Frankfurt, and smaller players like Azerbaijan taking their gold back to Baku.

The paper market is geared to certain depositories in New York and London and bank intermediaries that are members of the London Bullion Market Association. If you take the physical gold in New York and move it to Frankfurt, it reduces the floating supply available for leasing in New York. There is no well-developed leasing market in Frankfurt. The result of moving gold from New York to Frankfurt is to reduce the gold available to cover short positions in New York. That either increases systemic leverage or requires the short positions to be covered elsewhere in the physical market.

With regard to the pure physical market, there are a number of revealing transactions going on. If you're a large buyer of physical gold, you will discover that you have to source the gold directly from refineries, which means there are no sellers in the secondary market. In a normal, healthy market, if I'm a buyer, and someone else is a seller, a broker will find the two of us, buy from one, deliver to me, and earn a commission. Now what's happening is that there are buyers in the market but few sellers, so the only way a broker can get gold is straight from the refinery. The backlogs at refineries are running to about five or six weeks. That's how tight the physical supply situation is.

China, Russia, Iran, and other countries' central banks are stockpiling physical gold as quickly as they can. China is not doing this transparently. In July 2015, it updated its official gold reserves for the first time since 2009 to show 1,658 tons, up from the previous reserve of 1,054 tons. China has recently begun updating its gold reserves monthly to satisfy IMF reporting requirements. Yet China's figures are deceptively low because it holds huge gold reserves, perhaps 3,000 additional tons or more in an agency called the State Administration of Foreign Exchange, or SAFE. This gold is under the physical custody of the People's Liberation Army. Russia is more transparent. The Central Bank of Russia updates its gold reserve position monthly, and there is no evidence that it keeps large gold reserves off the books like China. Russia has about 1,400 tons in its reserves.

Russia can buy gold from its domestic mining production,

so it doesn't have to go to the market. China needs so much gold so quickly that even as the largest gold mining producer in the world, it's still not enough, so China purchases additional gold through the market. It does this through stealth, using covert operations and military assets in order to avoid the price impact of transparent market purchases.

International gold purchases by countries such as Russia, China, Iran, Turkey, Jordan, and others are accelerating. This lays a foundation for a massive global short squeeze in gold. However, this may not happen tomorrow: one should not underestimate the ability of central banks and major international banks to keep the gold manipulation game going longer than expected. If physical demand persists (and I expect it will), eventually the paper gold shorts are going to be squeezed, and the inverted pyramid of paper gold contracts will collapse. In the meantime, there's no doubt we are seeing price suppression through the paper gold market.

In 2013, I met in Switzerland with a senior executive of one of the largest gold refineries in the world. His factory is working flat out on triple shifts, twenty-four hours a day. He is selling all the gold he can produce, about twenty tons a week. Ten tons of that are going to China every week, a total of about five hundred tons a year from a single refinery. The Chinese want more. Still, he won't sell it to them, because he has to take care of other customers, such as Rolex, and other long-standing accounts that need gold for jewelry and watches.

He is sold out a year in advance and is having difficulty sourcing the gold.

The refinery takes gold from various sources, often old four-hundred-ounce bars, melts it down, re-refines it from 99.90 percent purity to 99.99 percent purity, casts it in the form of one-kilo bars, and delivers it to its customers, principally China.

Warehouses are being stripped bare. Periodically hundreds of tons of gold have come out of the GLD warehouse; COMEX warehouses are near all-time lows.

In Switzerland, I also met with vault operators, the secure logistics people who actually move and store the gold. They told me they couldn't build vault capacity fast enough. They are currently negotiating with the Swiss army to take over huge mountains in the Alps that are hollowed out and were used as military bases. Inside are tunnels and chambers formerly used by the Swiss army to store supplies, ammunition, and weapons. The Swiss army is abandoning some of these mountains and offering them to vault operators for gold storage. The secure logistics firms oversee the removal of gold from banks such as UBS, Credit Suisse, and Deutsche Bank and the entry into private storage in places like Brink's, Loomis, and other vault operators.

There are delays in the delivery of physical gold because the vaults and refiners are unable to keep up with the demand for both storage and newly refined gold. If physical gold is in such short supply, why has the gold price been under pressure in recent years? The answer is that massive shorting of futures and massive short sales of unallocated gold are continuing to keep the price down. The price of gold is a struggle akin to a tug-of-war between physical and paper transactions.

When you have a tug-of-war, it can be volatile. You have two powerful teams pulling against each other. You have the central banks, bullion banks, and hedge funds on one end of the rope and large acquirers and individual investors on the other. Sooner or later one team will give up or the rope will break. Any unexpected disruption in the delivery of physical gold could start a buying panic with skyrocketing prices. Such a disruption could be triggered by a failure to deliver gold as agreed, a gold exchange default, or the suicide of a prominent financier. These events are all well within the realm of possibility.

Rise in Price of Gold

In 2014, on a trip to Australia, I met with one of the largest gold bullion dealers in the country. Their CEO said their best months for sales came when the gold price was going down sharply. When small investors see the dollar price going down, they see a bargain. The Australian dealer told me that when the dollar price of gold dipped sharply, their customers were lined up out the door and around the block to buy gold.

I have consistently recommended that investors allocate about 10 percent of their investible assets to gold. This recommendation is intended as a long-term buy-and-hold position designed to preserve wealth in the event of sudden financial shocks and panics. Those pursuing this strategy don't become absorbed with daily price activity—it is what it is. The goal is not to make quick profits by trading the position. The goal is to preserve wealth for the long run. Having said that, it's obviously better to buy when the price dips rather than chase the upward spikes. Finding a good entry point is just common sense.

The extended decline in gold prices from 2011 to 2015 was discouraging to many investors. Yet, the recent price decline presented a buying opportunity for those who did not yet have a full 10 percent allocation.

There's a simple explanation for the price decline from the high point in 2011. Beginning in 2012, the dollar strengthened continually based on perceived and actual monetary tightening by the Fed. This included the beginning of "taper talk" in May 2013, the actual tapering of Fed money printing beginning in December 2013, removal of "forward guidance" by the Fed in March 2015, and continual talk about raising interest rates thereafter.

Over this period the euro collapsed from $1.40 to $1.05, and the Japanese yen also collapsed from about 90 to 120 yen per one dollar. More than fifty central banks around the world cut interest rates in 2015 to cheapen their currencies against the dollar. Prices on oil, sugar, coffee, and many other commodities collapsed from late 2014 to late 2015. Deflationary and disinflationary forces had the upper hand.

So you would expect the dollar price of gold to go down on a strong dollar. The question investors have to ask is, can that last? Is that the new state of the world? The answer is absolutely not. The reason is that the United States has allowed the dollar to appreciate and allowed other currencies to go down because these countries needed help. The Japanese economy was desperate to get inflation. The European economy had suffered its second recession inside the global depression that started in 2007. The United States allowed the dollar to go up, allowing the yen and the euro to go down, thereby throwing them a lifeline of monetary ease in the form of a cheaper currency.

The Fed's blunder was that the U.S. economy was itself not strong enough to bear the costs of a strong dollar. The Fed tightened into weakness in 2013, and with the usual monetary policy lags, disinflation appeared in the data by late 2014.

When you're worried about deflation, and you've cut interest rates to zero, printed trillions of dollars, and done everything else possible, the only way left to get inflation into your economy (which the Fed wants) is to cheapen the currency. Given the corner the Fed has painted itself into, my expectation is that the Fed will have to reverse course and again pursue monetary ease in the form of either more quantitative easing or a cheaper currency. Both paths are bullish for the dollar price of gold.

Manipulation

When we see the dollar price of gold go down sharply in the absence of relevant news, we can be reasonably certain that the gold market is being manipulated. There is statistical evidence, anecdotal evidence, and forensic evidence to support this conclusion. Gold manipulation is not new. Consider the 1960s London Gold Pool, or the U.S. and IMF gold dumping in the late 1970s. There is more recent evidence, including sales by the International Monetary Fund of four hundred tons of gold in 2010, that shows price suppression. Evidence of manipulation shows up in some recent academic studies as well. The manipulation is real.

If your goal as a central bank is to keep the price of gold from moving in a disorderly way, you have to manipulate the market only when gold's going up. If it's going down for more fundamental reasons such as deflation, then if you're a central bank that wants a weak gold price, you're getting what you want and no added manipulation is needed. Manipulation really kicks in when gold is strong and looks like it's breaking out. We saw that in August 2011 when it was rapidly approaching $2,000 per ounce. That was an important psychological barrier, and gold could have gone up a lot higher from there, so the central banks really did have to use extraordinary effort to drive the price down.

Let's look specifically at manipulation techniques.

Dumping Physical Gold

The most straightforward technique for suppressing the dollar price of gold, and the most obvious, is to just dump physical gold. If you're a central bank, sell gold. That was done for decades, starting with the 1960s London Gold Pool, when members of the Bretton Woods system including West Germany, the United States, and the United Kingdom took turns dumping gold on the London bullion market to suppress the price.

Such efforts continued in the 1970s after Nixon went off the gold standard. Gold started that decade at $35 per ounce. After Nixon shut the gold window, it went up to about $42 per ounce. By January 1980 it hit $800 per ounce. On the path from $42 an ounce to $800 an ounce, the United States tried desperately and secretly to suppress the gold price with gold sales. (This effort is described in detail in chapters 9 and 11 of my book *The Death of Money*.)

Between 1974 and 1980, the United States sold about 1,000 tons and prevailed upon the IMF to sell 700 tons, so together the United States and the IMF dumped 1,700 tons of physical gold, about 5 percent of all the official gold in the world. This manipulation effort failed. The dollar price of gold skyrocketed to $800 per ounce by January 1980 despite the physical gold dumping on the market. So at the end of the day, the United States just gave up and let the gold price go where it was going to go.

I've found classified, private correspondence (since declassified) from the mid-1970s among Arthur Burns, chairman of the Federal Reserve at the time, U.S. president Gerald Ford, and the chancellor of West Germany all describing this physical gold manipulation.

There was more manipulation all the way through the late 1990s, including the infamous "Brown's Bottom," when Gordon Brown, then chancellor of the British exchequer, dumped about two thirds of the UK's gold on the market in 1999 at close to the lowest price of the past thirty-five years.

Switzerland was also a major gold seller in the early 2000s. So for a long time, the way major financial powers manipulated the price was by dumping physical gold. The problem is that eventually you start to run out of gold. The UK doesn't have much left. Switzerland has a fair amount yet a lot less than it used to. The United States decided it didn't want to sell any more gold yet was happy for others to sell. The physical dumping eventually stopped because people realized it wasn't working. There were ready buyers, and the manipulators were running low on gold. They had to resort to paper manipulation. Let's talk about how that works.

Paper Manipulation

The easiest way to perform paper manipulation is through COMEX futures. Rigging futures markets is child's play. You just wait until a little bit before the close and put in a massive sell order. By doing this you scare the other side of the market into lowering their bid price; they back away. That lower price then gets trumpeted around the world as the "price" of gold, discouraging investors and hurting sentiment. The price decline spooks hedge funds into dumping more gold as they hit "stop-loss" limits on their positions. A self-fulfilling momentum is established where selling begets more selling and the price spirals down for no particular reason except that someone wanted it that way. Eventually a bottom is established and buyers step in, but by then the damage is done.

Futures have a huge amount of leverage that can easily reach 20 to 1. For $10 million of cash margin, I can sell $200 million of paper gold. We know who the brokers on the floor and the clearing brokers are, but the market is still nontransparent, because we don't know who the players actually are—those doing the buying or selling through the brokers. We don't know who the ultimate customer is. Only the brokers know that, so there is anonymity through the broker combined with high leverage.

Paper manipulation can also be performed through exchange-traded funds, or ETFs, notably GLD. Market manipulations using GLD are more complicated. The GLD ETF is actually a share of stock. The stock is in a trust, which takes your money, buys gold, and puts it in a vault. If you don't like gold or the price action, all you can do is sell your shares.

It's entirely possible to have physical gold trading one way and ETF shares trading another way, thereby opening up a "spread" between the two prices. They should be closely aligned. Still, such a spread or arbitrage does exist from time to time.

Here's what happens. If I'm one of the big banks that are authorized participants in GLD, I look for the arbitrage. I see the physical gold price trade at a premium to the shares (the share equates to a certain amount of gold). I short the physical gold and simultaneously buy the shares from somebody in the market who's been spooked out of it. Now I take the shares to the trustee and cash them in to get the physical gold. I deliver the physical gold to cover my physical short, and I keep the difference. It's an (almost) risk-free arbitrage.

One result of such activity is to take the gold out of the ETF warehouse, reducing the floating supply. The floating supply is the gold available for trading to support the paper gold. If gold is in a bullion bank or a GLD or COMEX warehouse, it's part of the floating supply and available to support paper trading. Once it gets to China and goes into a Shanghai vault or it gets to a Loomis vault in Switzerland it's no longer part of the floating supply. It is a part of the total stock, yet it's not available for trading or other transactions such as leasing and forward sales.

Importantly, the gold in Fort Knox or at the Federal

Reserve Bank of New York or COMEX may be leased or leveraged, if it's not being sold outright. The Chinese are buying gold that's not going to see the light of day for an indefinite period because they're putting it in deep storage. Once it gets to China, it's not coming back out again. The Chinese are not day traders or flippers. They're buying enormous amounts of gold and putting it in storage where it will stay.

Taking all of these flows into account reveals that more and more paper gold trading is resting on less and less physical gold. The inverted pyramid of paper gold contracts is resting on a small base of physical gold, and that base keeps getting smaller as the Russians and Chinese accumulate gold.

Hedge Fund Manipulation

Hedge funds are now large players in the gold market. Historically, that was not the case. Gold ownership distribution looked something like a barbell. At one end, you had the small holder who always felt more comfortable with gold coins or bars in her possession. On the other end were the largest holders, sovereign wealth funds and central banks. In between, you didn't really see institutions much involved one way or the other. That's less true today. Hedge funds are beginning to fill in the middle ground of gold investors between retail and the sovereigns.

To a hedge fund, gold may be an interesting market in which

to deploy its trading style. Still, gold is not special; it's just another tradable commodity. To hedge funds, the commodity could just as well be coffee beans, soybeans, Treasury bonds, or any other traded good.

Hedge funds use what are called "stop-loss" limits. When they establish a trading position, they set a maximum amount they are willing to lose before they get out. Once that limit is reached, they automatically sell the position regardless of their long-term view of the metal. Perhaps they don't even have a long-term view, just a short-term trading perspective.

If a particular hedge fund wants to manipulate the gold market from the short side, all it has to do is throw in a large sell order, push gold down a certain amount, and once it hits that amount, these stops are triggered at the funds that are long gold. Once one hedge fund hits a stop-loss price, that hedge fund automatically sells. That drives the price down more. The next hedge fund hits its stop-loss. Then it sells too, driving the price down again. Selling gathers momentum, and soon everyone is selling.

Eventually the price can work its way higher again, more funds will begin to acquire gold, and then the short-side manipulators get to play the game all over again, driving the price lower time after time. In the absence of government enforcement of antimanipulation rules, gold holders should expect these games to continue until a fundamental development drives the price to a permanently higher plateau.

Leasing Unallocated Forwards

Another way to manipulate the price is through gold leasing and unallocated forwards. "Unallocated" is one of those buzzwords in the gold market. When most large gold buyers want to buy physical gold, they'll call JPMorgan Chase, HSBC, Citibank, or one of the large gold dealers. They'll put in an order for, say, $5 million worth of gold, about five thousand ounces at market prices as of this writing.

The bank will say fine, send us your money for the gold and we'll offer you a written contract in a standard form. Yet if you read the contract, it says you own gold on an "unallocated" basis. That means you don't have designated bars. There's no group of gold bars that have your name on them or specific gold bar serial numbers that are registered to you. In practice, unallocated gold allows the bank to sell the same physical gold ten times over to ten different buyers.

It's no different from any other kind of fractional reserve banking. Banks never have as much cash on hand as they do deposits. Every depositor in a bank thinks he can walk in and get cash whenever he wants, but every banker knows the bank doesn't have that much cash. The bank puts the money out on loan or buys securities; banks are highly leveraged institutions. If everyone showed up for the cash at once, there's no way the bank could pay it. That's why the lender of last resort, the Federal Reserve, can just print the money if need be. It's no different in the physical gold market, except there is no gold lender of last resort.

Banks sell more gold than they have. If every holder of unallocated gold showed up all at once and said, "Please give me my gold," there wouldn't be nearly enough to go around. Yet people don't want the physical gold for the most part. There are risks involved, storage costs, transportation costs, and insurance costs. They're happy to leave it in the bank. What they may not realize is that the bank doesn't actually have it either.

A central bank can lease gold to one of the London Bullion Market Association (LBMA) banks, which include large players like Goldman Sachs, Citibank, JPMorgan Chase, and HSBC. Gold leasing is often conducted through an unaccountable intermediary called the Bank for International Settlements (BIS). Historically, the BIS has been used as a major channel for manipulating the gold market and for conducting sales of gold between central banks and commercial banks.

The BIS can take the gold it already leased from the Federal Reserve and lease it to commercial banks that are LBMA members. The commercial banks then have title to a certain amount of physical gold. They then sell ten times as much to the marketplace on an unallocated basis. So you can see the leverage at work there. They can sell as much gold as they want and they don't need to have *any* physical gold, just a paper title through the lease agreement.

None of this is speculation. You can go to the BIS annual

report and look in the footnotes where it discloses the existence of lease arrangements with central banks and commercial banks. It doesn't mention the banks by name, but the activity itself is clearly disclosed. We know who the commercial banks are because they have to be LBMA members, and we know who the central bank lessors are, so there's no need for speculation about what is going on.

The BIS, based in Basel, Switzerland, has an intriguing and somewhat checkered history. It was founded in 1930 as a result of efforts throughout the 1920s led by the Bank of England. It can be seen as a Swiss tree house where children hang out without supervision or scrutiny, except instead of children you have central bankers acting without supervision or scrutiny.

Once a month, the major central bankers in the world gather in Basel, organized in tiers. There's a larger group of as many as fifty BIS members and there's an inner group of seven to ten or so, a relatively small number of central bankers.

The larger group gathers for certain meetings, but the inner group of ten go off on their own, shut the doors, and make their own deals. BIS is the most nontransparent institution in the world. Even intelligence agencies such as the CIA suffer occasional leaks, yet BIS leaks are unheard of. They don't put much on the Web site. They do a lot of technical research you can access and they actually do have audited financial statements. Still, they don't tell you about their deliberations. There are no minutes released of what goes on behind closed doors, and no press conferences after the central bankers meet. BIS is the ideal venue for central banks to manipulate the global financial markets, including gold, with complete nontransparency.

Combining Manipulations

You can combine manipulations. Let's start with an LBMA bank dealer. That dealer sees there's demand for physical gold in China, and there's gold in the GLD warehouse in London. Here's what it can do. First go into the futures market and slam the gold price. This spooks the little guy who starts selling his GLD shares, driving down the share price. Meanwhile, the smart money sees a good entry point. The little guy is dumping his GLD shares while the big players are buying physical gold. Momentum opens up the spread between physical and GLD shares.

The LBMA dealer next sells physical gold short to China. The dealer then buys shares from the little guy who's scared to death. The dealer redeems the shares, gets the physical gold, delivers it to China, and pockets the difference. So a dealer can create its own supply, create its own arbitrage, and then profit from the difference. This type of manipulation arbitrage took place in 2013 when the GLD warehouse disgorged five hundred tons of gold as the gold price went down for the first time in a dozen years. In the course of these manipulations, the floating supply was reduced and more physical gold ended up in China.

Who's Behind the Manipulation?

We've looked at how manipulation works, how it used to be done in the physical market, and how it's done today, primarily through COMEX, ETFs, hedge funds, and leasing and unallocated contracts.

The next questions are: "Why?" And "Who's behind the manipulation?" The LBMA banks are in it for the arbitrage and dealing profits and the hedge funds are in it for the momentum profits. Yet are there larger political and policy interests involved? There are two players in the world with a strong motive to suppress gold prices, at least in the short run: one is the United States, and the other is China.

Many observers have a naïve and, in my view, mistaken analysis of the Fed's interest here. Observers assume the Fed wants to squash the gold price to give an impression of dollar strength. In reality the Fed wants a weaker dollar because it's desperate for inflation. It doesn't want the dollar to go away or collapse, yet a cheaper dollar will cause imports to cost more, which helps the Fed to meet its inflation targets. The United States is a net importer. A cheaper dollar means import prices go up, and inflation feeds through the supply chain in the United States.

A weaker dollar should mean a higher dollar price for gold. Still, there are two constraints on a weak dollar/strong gold hypothesis. The first constraint is that just because the Fed wants a weaker dollar does not mean it automatically gets it. There are countervailing forces, including the natural deflationary tendencies from demographics, technology, debt, and deleveraging. The other countervailing force is the fact that other countries also want weak currencies to help their own economies. Retaliation is the root dynamic of currency wars. Because two currencies cannot devalue against each other at the same time, the need for a weak yen or euro to help Japan or Europe necessarily implies a stronger dollar (and weaker gold), even if the Fed wants the opposite. However, in the long run the Fed does not mind a weak dollar/strong gold policy.

There is a condition on any long-run policy of higher gold prices. It must be *orderly*, not disorderly in the Fed's perspective. Slow, steady increases in the dollar price of gold are not a problem for the Fed. What the Fed fears are huge, disorderly moves of $100 per ounce per day that seem to gather upward momentum. When that happens, the Fed will immediately take steps to rein in the upward price momentum. Whether those steps will succeed or not remains to be seen.

A good example is the July, August, and early September 2011 period. At that time the gold price was skyrocketing.

It went up from $1,700 to $1,900 per ounce quickly and was clearly headed for $2,000 an ounce. Once you get to $2,000 per ounce the momentum psychology can feed on itself. The next stop could have been $3,000 per ounce; clearly a disorderly process.

The gold price action was getting out of control. The Fed manipulated the price lower, not because it ultimately wanted a lower price, but because it was worried about a disorderly increase. The Fed is perfectly fine with an orderly increase as long as it doesn't go up too far, too fast, and change inflationary expectations. The Fed will be in the market to manipulate when it finds it necessary.

Now let's consider the other major player—China. China definitely wants a lower price because it's buying. It sounds like a paradox—China owns a lot of gold; why would it want the price to go down? The reason is that it's not done buying. China probably needs several thousand more tons of gold before it catches up to the United States. It's precisely because China is still buying that it wants the price to stay low. This gives China a motivation to manipulate the gold price.

The interaction of these U.S. and Chinese preferences has interesting policy implications. The U.S. Treasury to some extent needs to accommodate Chinese wishes because China owns several trillion dollars of U.S. Treasury notes. While the Fed and the Treasury want inflation to help manage the U.S. debt load, China fears that inflation will erode the value of its Treasury holdings.

If inflation breaks out, China's incentive is to dump Treasuries, which would raise interest rates in the United States and sink the U.S. stock market and housing market.

The compromise between the Fed's desire for inflation and China's desire to protect its reserves is for China to buy cheap gold. That way, if inflation is low, China's gold won't go up much, but the value of its paper Treasury reserves is preserved. If the United States gets the inflation it wants, China's Treasuries will be worth less, yet its gold will be worth much more. Having Treasuries and gold is a hedged position that protects China's wealth even as the Treasury tries to destroy U.S. savers' wealth with inflation. The solution for U.S. savers is to do exactly what China has done—buy gold.

Contrary to much speculation, China is *not* buying gold to launch a gold-backed currency, at least not in the short run, but to hedge its Treasury position. The Treasury has to accommodate that or else China will reduce its position in Treasuries.

What remains is a strange condominium of interests where the Treasury and China are in agreement that China needs more gold and the price cannot be too high or else China could not easily afford all it needs. This is an issue I have discussed with senior officials at the IMF and the Fed, and they've confirmed my understanding that a global rebalancing of gold from the West to the East needs to proceed, albeit in an orderly way.

The United States is letting China manipulate the market so China can buy gold more cheaply. The Fed occasionally manipulates the market as well so that any price rise isn't disorderly. Where does the manipulation end? What can individual investors do to weather the gathering storm?

Beating the Manipulation

When we hear about the enormous forces brought to bear on the gold price—with the United States on the one hand and China on the other—how does the individual investor stand up against such forces?

There is an inclination to say: "I can't win against these players, therefore it's not worth the risk of being in the gold market." In the short run, it's correct that you can't beat them, but in the long run, you always will, because these manipulations have a finite life. Eventually the manipulators run out of physical gold, or a change in inflation expectations leads to price surges even governments cannot control. There is an endgame.

History shows manipulations can last for a long time yet always fail in the end. They failed in the 1960s London Gold Pool, with the United States dumping in the late 1970s, and the central bank dumping in the 1990s and early 2000s. The gold price went relentlessly higher from $35 per ounce in 1968 when the London Gold Pool failed to $1,900 per ounce in 2011, the all-time high. There are new forms of manipulation going on now, yet ultimately they always fail. The dollar price of gold will resume its march higher.

The other weakness in the manipulation schemes appears in the use of paper gold through leasing, hedge funds, and unallocated gold forwards. These techniques are powerful. Still, any manipulation requires *some* physical gold. It may not be a lot, perhaps less than one percent of all the paper transactions, yet some physical gold is needed. The physical gold is also rapidly disappearing as more countries are buying it up. That puts a limit on the amount of paper gold transactions that can be implemented.

For example, the manipulation that took place in 2013, when the GLD warehouse disgorged five hundred tons of gold, could not be replicated because by 2014 there were only about eight hundred tons of gold left in GLD. If GLD were to disgorge another five hundred tons, there would not be enough gold left in GLD to make the ETF financially viable for the sponsor. There comes a time when the amount of gold left is so small that the management fees don't cover the costs of insurance, storage, administration, and other expenses.

The third point to consider is that there is an endgame that arrives when China has enough gold so that its gold-to-GDP ratio equals or exceeds that of the United States. It's not there, yet once it is, there will be no political reason to buy more. China will have secured an equal voice at the table the next time a Bretton Woods–style conference is needed to restore confidence in the international monetary system.

Once China has enough gold, the United States and China together could let the gold price go wherever it wants in an orderly way. Inflation could get out of control, and China wouldn't lose. If inflation and the gold price skyrocketed right now, China would be left in the dust. It doesn't have enough gold to hedge the portfolio losses on its Treasury holdings. With the gold price soaring and the Chinese economy growing faster than that of the United States, China would never catch up in hitting a gold-to-GDP target.

China is buying as much gold as it can, but because it's trying to target a gold-to-GDP ratio and has the fastest-growing major economy in the world, it's a moving target. The price has to be kept down until China has enough gold. When it's done buying, when it has approximately eight thousand tons, the United States and China can shake hands and both say they're protected. At that point, a dollar devaluation by a rise in the dollar price of gold can commence.

My advice to investors is that it's important to understand the dynamics behind gold pricing. You need to understand how the manipulation works, what the endgame is, and what the physical supply-demand picture looks like. Understanding these dynamics lets you see the endgame more clearly and supports the rationale for owning gold even when short-term price movements are adverse.

Chapter 5

GOLD IS RESILIENT

Gold traded down to levels in the $1,150 to $1,050 range four times between 2013 and 2016. Every time, however, it bounced back. Gold's value has shown resilience in a highly adverse environment. Many investors are frustrated that gold hasn't gone higher. Still, one might feel encouraged that it didn't trade lower, considering what has happened to commodity prices in general, and the rise of real interest rates as inflation declined. We are living in a massively deflationary world. Gold has bounced off the bottom multiple times and shown relative strength. That's a good sign going forward.

Gold has maintained its resilience through monetary collapses in the past, and it will do so in future collapses. This will be true especially in the face of a new and powerful threat: cyberfinancial wars.

Cyberfinancial Wars

On August 22, 2013, the NASDAQ was shut down for half a day. Investors have never been given a credible explanation as to what happened. If there were a benign or technical explanation, NASDAQ would have told us about it by now. They could have said there was a bad piece of code or an engineer blundered while updating software or an installation didn't go well. NASDAQ has never provided information of any substance except a few vague references to an "interface problem."

Why not? NASDAQ itself must know. One likely answer is that the cause of the shutdown was nefarious, and it was probably caused by criminal hackers or, worse yet, Chinese or Russian military cyberbrigades. Investors should have no doubt about the ability of a number of foreign cyberwarfare units to close or disrupt major stock exchanges in the United States and elsewhere.

In 2014, *Bloomberg Businessweek* broke a story with a cover article titled "The NASDAQ Hack." The incident referred to in the title goes back to 2010. Yet it was only in late July 2014 that the media were able to report on what happened: with help from the FBI, NSA, and Department of Homeland Security, the NASDAQ actually found a computer virus in its operating systems, traced it back to its source, and determined it was an attack virus. It wasn't put there by a criminal gang; it was planted by the Russian state.

Stories of this type are often served up to reporters from official sources with an agenda. Why did this particular story come out four years after the incident? The reporting is timely, but why did the source wait four years? One surmise is that an administration official wanted to reveal the extent of the Russian invasion of U.S. financial exchanges as a way to alert investors to the possibility of worse to come. It was a warning.

A common response from analysts is that our hackers must be as good as theirs; we could close down the Moscow Exchange if Russian hackers were to close down the New York Stock Exchange. Yes, of course we could. The United States is actually better at cyberwar than any other world power. But consider how that would play out.

If Russia shuts down the New York Stock Exchange and we shut down the Moscow Exchange, who loses? We lose, because our markets are more important and much larger. There's far more wealth involved on our side and greater spillover effects. Russia, financially, is in the position of not having as much to lose. One reason to avoid retaliation and escalation in cyberwarfare is because it ends badly for the United States. Russian president Vladimir Putin knows that too, and that's one of the reasons he invaded Crimea with confidence in 2014. He knew perfectly well the United States could not escalate in the financial battle because, in the end, we had more to lose than Russia.

For those unfamiliar with the Cold War, an escalation dynamic existed then also. The United States had enough missiles to completely destroy Russia (then called the Soviet Union). Russia had enough missiles to completely destroy the United States. This is a highly unstable situation because there was a great temptation to launch first. If you strike first and wipe out the other guy, you win. The response to this instability was to build more missiles. With enough missiles you could withstand the first strike and still have enough left over to launch a second strike. The second strike would devastate the party that started the war in the first place. It was that second-strike capability that prevented the other player from launching his missiles first.

This same dynamic as applied to financial warfare is not fully appreciated today. The weapons may be symmetric but the losses are not. The United States has by far the most to lose.

Another danger is the accidental launching of a cyberfinancial war. If you ask your hackers to devise an ability to shut down the New York Stock Exchange, they have to practice that. They have to launch probes. For example, a situation could arise where Russian hackers who don't intend to start a financial panic are probing and accidentally start a financial panic or an exchange systems shutdown. That is the much more worrisome scenario, because it doesn't require irrationality. It only requires an accident, and of course, accidents happen all the time.

The United States has excellent deterrent capabilities in

cyberwarfare through the military's Cyber Command and the National Security Agency (NSA). However, insufficient effort has been devoted to strategic doctrine. Only a few experts such as Juan Zarate at the Center on Sanctions and Illicit Finance, and Jim Lewis at the Center for Strategic and International Studies are performing roles comparable to those that Herman Kahn and Henry Kissinger performed in the 1960s when strategic nuclear war fighting doctrine evolved. This strategic deficiency increases the risk of cyberfinancial war. That threat is one more reason to own gold because it is not digital and cannot be hacked or erased.

Abandoning the Dollar

Though it seems like an extraordinary policy to adopt, since 2010 the U.S. government has effectively abandoned the sound dollar. In January of that year, the United States ended the sound dollar policy that had prevailed since 1980. An intentional policy of cheapening the dollar to encourage inflation and nominal growth was commenced. The policy had been worked out at the G20 Leaders' Summit in Pittsburgh held in September 2009. The view was that the United States was the world's largest economy and if growth in the United States collapsed it would take the world down with it. A cheap dollar was the key to growth, so the sound dollar was abandoned.

The cheap dollar tactic started a currency war that has been playing out ever since. One problem with currency wars is that they have no logical conclusion. In the case of the dollar, there's a lot of activity around the world designed to diminish the dollar's role as a global reserve currency. Too many of our trading partners and financial investment partners have lost confidence in the dollar and resent the way the United States uses the dollar's status to run deficits and print money to cover the gap.

For example, there was an enormous uproar in France in 2014 when the United States extracted almost $9 billion from BNP Paribas, one of the largest French banks, for violating U.S. economic sanctions. The actions that comprised the violation took place in France, Switzerland, and Iran—completely outside of U.S. jurisdiction—and were committed by French banks and Iranian counterparties. But because the transactions were denominated in dollars, and those dollars had to flow through a settlement system controlled by the Federal Reserve and the U.S. Treasury, they became subject to U.S. jurisdiction even though prima facie there is nothing in the transaction itself to subject those banks to U.S. law.

You can argue the merits of such prosecutions both ways. Still, there's no doubt that trading partners, including allies of the United States, are fed up with the global dollar system partly because of such prosecutions. As a result, foreign banks are moving away from the dollar system as quickly as they can.

Dollar Hegemony

Trade finance between two sovereign countries is just a matter of keeping score. For example, if I ship goods to you and you owe me in one currency, and you ship goods to me and I owe you in another currency, the two foreign currency balances can be netted, with the net settled in any currency of our choice. That's the balance of trade, and it gets settled up periodically. You can count that in dollars, baseball cards, or bottle caps. Any *numéraire* or unit of account the parties agree on is sufficient for that purpose. This means that a broad array of currencies can serve as trade currencies. The Chinese yuan certainly qualifies.

There's a difference between a trade currency and a reserve currency. The reserve currency is not just a way of settling trade balances. It's how you invest your surplus. To act as a reserve currency requires deep, liquid pools of investible assets. That's why the Chinese yuan is not close to real reserve currency status—the pool of investible assets just isn't there.

There is no market in the world that comes close to the U.S. Treasury market as a place that can absorb the capital flows that result from world trade and investment. When you consider the reserve positions of Japan, China, Taiwan, and a few others, the amounts are measured in the trillions of dollars. In the short run, there is no market capable of absorbing these capital flows in liquid form except the U.S. Treasury securities market.

Having said that, there is no doubt that Russia, China, and others would like to get out from under the dollar hegemony. They would like a system that is not based on the dollar, yet there are hurdles in the way.

China is fearful because it has $2 trillion of U.S.-dollar-denominated debt in its $3.2 trillion of reserve positions (the rest is in gold, euros, and other assets), and it's worried the United States is going to inflate the currency.

Russia wants to turn its back on the dollar because we opposed its ambitions in Eastern Europe and Central Asia in 2015 and imposed dollar- and euro-based sanctions.

Saudi Arabia may want to turn its back on the dollar because it feels betrayed by the United States. In December 2013, President Obama effectively anointed Iran as the regional hegemonic power in the Persian Gulf, allowing Iran to keep its nuclear reactor and its uranium enrichment program. That's tantamount to recognizing Iran as the leading regional power. Saudi Arabia saw this as a stab in the back, particularly because of a secret U.S. agreement from decades prior.

In the 1970s, during the Nixon and Ford administrations, the United States and Saudi Arabia agreed on the petrodollar deal. The United States would ensure Saudi Arabia's national security, and in exchange Saudi Arabia would require that oil be priced in dollars. Once oil was priced in dollars, the entire world would need dollars because everyone needs oil. The petrodollar agreement created a strong foundation for keeping the dollar as the global reserve currency.

Today China, Russia, and Saudi Arabia, all powerful nations that export oil, natural gas, and manufactured goods, have a shared interest in ending dollar hegemony in the international monetary system.

In 2009, I was one of the game designers and facilitators of the Pentagon's first-ever financial war game, an event I wrote about in my first book, *Currency Wars* (2011). When the game commenced, I played on the China team and my colleague played on the Russia team. Together we devised a plan where Russia and China would combine their gold reserves, place them in a Swiss vault, and issue a new currency from a London bank backed by that gold. China and Russia would then announce that, going forward, if you wanted to buy Russian energy or Chinese manufactured goods, they would no longer take dollars. They would only take their new currency. If you wanted the new currency you could earn it, borrow it, or you could deposit your gold side by side with the Russian and Chinese gold and obtain the new currency from the London bank. The bank would issue it backed by gold.

So all of a sudden, you had a new gold standard, a new currency sponsored by Russia and China with others invited to participate, and an urgent need to use it, because you would have to use it to get Russian and Chinese exports.

We knew the Russia-China gold scenario wouldn't play

out immediately. The whole point of a war game is to do some out-of-the-box thinking and help the Pentagon look far into the future—to look over the next ridgeline.

At the time, we were ridiculed by some of our fellow participants who are widely respected economists. The supposed gurus insisted we were being ridiculous and that gold is not part of the international monetary system. We were accused of wasting everyone's time.

That's fine, we thought; we'll just grin and bear it and see how the game plays out. Still, since we devised this scenario in 2009, Russia has increased its gold reserves by 100 percent and China has increased its gold reserves several hundred percent. In other words, China and Russia are behaving exactly the way we predicted they would behave. They can see the crack-up of the international monetary system coming. They're preparing for it by acquiring gold. Investors should do the same.

This does not mean that tomorrow morning we wake up and the ruble is a gold-backed global reserve currency. I don't expect that at all. There are many problems with Russian corruption, Russian rule of law, and the absence of a significant Russian bond market. We won't see the ruble as a global reserve currency anytime soon. Yet Russia and China are nevertheless moving away from the dollar and toward gold.

We saw evidence of this movement in July 2014 with the announcement of the China-Russia multiyear, multibillion-dollar natural gas and oil trade and development deals. Russia then announced a similar, somewhat smaller yet still quite large deal with Iran. Both Iran and Russia are undergoing economic sanctions from the United States. Iran was actually kicked out of the dollar payment system for a time. That hasn't happened yet with Russia, yet the United States threatens it at times. Iran and Russia are also joining hands to escape the dollar trap with new deals involving weapons, nuclear reactors, gold, and food.

What's interesting is that Russia agreed to buy Iranian oil. That's odd, because Russia is one of the largest oil exporters in the world. Why would Russia import oil from Iran if it is a major exporter on its own? The answer is that until recently the Iranians could not easily sell their oil on the open market because of U.S. sanctions. If Iran sells the oil to Russia, it can be reexported by the Russians (oil is fungible to an extent) to the Chinese and others. Russia, already under U.S. sanctions, can act as a middleman in the sale of oil by Iran to China (also under U.S. sanctions) and keep Chinese hands clean.

Recently, China did a currency swap with Switzerland so China can access Swiss francs in exchange for yuan. Now we can start to connect the dots. China has access to Swiss francs, a highly desirable hard currency. Iran is selling oil to Russia, which Russia can resell to China. China can pay Russia in Swiss francs that Russia can intermediate through the new BRICS bank. What's missing in this chain of commerce? The dollar's missing—it's not involved.

All of these countries are working behind the scenes to end the dollar hegemony while the United States seems to be asleep at the switch. Investors will wake up one day and find that the dollar is in free fall and they won't know why. Still, you can already see this big switch coming. If the dollar collapses from a lack of confidence, the entire international monetary system collapses too. That is what I expect.

The Role of Emerging Markets

Current U.S. monetary policy is having deleterious effects on emerging markets, and these markets are powerless to overcome these effects except by moving interest rates up or down or imposing capital controls. This is how markets work in a "world without an anchor."

The Fed has tried to wash its hands of emerging markets. Federal Reserve officials including Ben Bernanke and Janet Yellen have repeatedly said that their job is to focus on U.S. economic performance. They say it's not the Fed's job to worry about emerging markets. From the Fed's perspective, emerging markets are collateral damage in the currency wars. The Fed behaves like a drunk driver who runs down pedestrians, and then blames the pedestrians for being in the way.

For example, the Federal Reserve implicitly says to South Africa: "If you believe your currency is too weak, then raise interest rates." Well, how can South Africa raise interest rates without exacerbating its serious unemployment problems? These conundrums are true around the world. The Fed is being disingenuous about ignoring the impact of U.S. monetary policy on the rest of the world.

The dollar is still the world's dominant reserve currency, at least for the time being. The emerging markets, the BRICS, have most of their reserves in dollars. The dollar capital markets are still huge in comparison with these emerging markets. As a result, emerging markets are highly vulnerable to hot money capital inflows and outflows.

When dollar capital flows in and out of these emerging markets, based on the Fed's market manipulation, it can overwhelm them. Money floods into emerging markets when the Fed supports a "risk on" mode. This same capital can flow out just as quickly when the Fed talks tough on interest rates and the world is in a "risk off" mode. From the emerging markets' perspective, it's practically an invitation to impose capital controls. Many of the central banks in these emerging markets, including those of Jordan, Malaysia, the Philippines, and Vietnam, have been increasing their gold reserves in recent years as a hedge against dollar instability.

One of the dangers the Federal Reserve is facing is that it may trigger an emerging market crisis. Emerging markets don't know which way to turn, because they're dependent on the dollar. The Fed is manipulating the dollar through interest rate policy, which means it's indirectly manipulating every market in the world.

With these emerging markets so vulnerable, it wouldn't surprise me to see some of them putting on capital controls. For twenty years, we've talked endlessly about globalization, and the integration of capital markets across borders. Today these markets are densely connected. Globalization on the way up means globalization on the way down. A serious balance-of-payments or reserve crisis in any of the leading emerging markets' economies will quickly spiral out of control.

This is what happened in 1997 and 1998 and it almost collapsed capital markets around the world. That particular collapse started in Thailand, spread to Indonesia and South Korea, and then finally to Russia. A resulting Russian default led to the collapse of Long-Term Capital Management. I was at LTCM at the time, so I had a front-row seat on that one. It came close to crashing every stock and bond market in the world until the Fed and IMF intervened. That dynamic can easily happen again.

Chaos and Collapse

Investors and everyday citizens know there is latent instability in our global monetary system. I've heard such sentiment from people in countries all over the world. Still, those in power centers like the Federal Reserve, the U.S. Treasury, and the IMF are slow to acknowledge the problems in the system. The global monetary system is heading over the cliff, yet it's difficult to find a good appreciation of the problem in policy circles, or any willingness to embrace solutions.

We are getting closer to a collapse of the international monetary system. That doesn't mean tomorrow morning necessarily, but it does mean that it will happen sooner rather than later. It's not a ten-year forecast. Could it be five years? Maybe. Could it be one year? Yes.

We can never know the exact time the collapse will take place, yet it's somewhere in that medium-term time frame and soon enough to begin taking action today. An international monetary crisis doesn't mean you automatically go to a gold standard. Still, that is one of the possible outcomes that may be necessary to restore confidence.

If you do revert to a gold standard, you have to determine what the dollar price of gold is going to be. One of the great economic blunders of the twentieth century—and maybe of all time—occurred when nations went back to a gold standard in the 1920s at the wrong price. They had printed so much money to finance World War I that going back to the gold standard at the prewar price was disastrously deflationary. It was necessary in the United Kingdom to shrink the money supply to reestablish the old gold-money parity. The United Kingdom should have simply recognized the fact that it had printed the money, and relaunched the gold at a much higher price per ounce.

If you were going back to a gold standard today, it would be critical to avoid this blunder of the 1920s. The math is straightforward. To have a gold standard today that was non-deflationary, you'd have to have a price between $10,000 an ounce and $50,000 an ounce depending on which assumptions you want to make about the choice of money supply, the percentage of gold backing, and the specific countries that would be included in the new system. I'm not predicting or expecting $50,000 gold, yet I do expect $10,000 per ounce of gold based on a return to some sort of gold standard.

If governments around the world were taking steps to avoid the monetary collapse, I might change my forecast and conclude we're steering away from disaster because we're enacting smart policy. In fact, I see the opposite. I don't see smart policies, I see the collapse coming and leading to a much higher dollar price for gold in order to restore global confidence. These are not numbers I make up to be provocative or get headlines. It's a straightforward analysis based on readily available data.

Having said that, gold has indeed been volatile. It's not a lot of fun if you're a gold investor watching the dollar price go down; no one enjoys that. Personally, I don't get too euphoric when the dollar price goes up, and I don't get too depressed when the dollar price goes down. To me, it's just market information that provides me with insights into the underlying dynamics. When I see the dollar price going down, I certainly don't sell gold. Occasionally, I buy more because I like the price and think it's probably a good entry point.

The most likely outcome from the dollar's fall is chaos or collapse emerging from the complexity of the global financial system. It's not what anybody wants. I don't think there's any team out there rooting for chaos and collapse.

Yet it will happen because of the dynamic instability of the system, the inability to analyze the risks correctly, wishful thinking, denial, delay, and the application of bad science by economists. Many of these analytic deficiencies arise from certain cognitive traits that are part of human nature. Collapse is not what anyone wants, but we may well get collapse, because policymakers, such as central bank heads, finance ministers, IMF officials, heads of state, and the G20 leaders, misapprehend market risks and are not taking sufficient steps to change the system. So I see chaos as the most likely outcome.

If you're an investor, or a portfolio manager, or you're just trying to make sense of it all, I don't think it serves you well to make categorical predictions about the timing and catalyst for a collapse. What I try to do instead is identify what I call indications and warnings (we do this in intelligence analysis).

As an illustrative problem-solving exercise, assume there are four paths from where I stand today. I set out on a journey. I don't know which path I'm on although I may form a guess based on whatever scraps of data are at hand. However, there are signposts along the way. As I read the signposts, it helps me to understand which path I'm on. For example: I live in the New York area. On the way to Boston, the roadside restaurants are McDonald's, and on the way to Philadelphia, the roadside restaurants are Burger Kings. If I see a Burger King, I know I'm not going to Boston, so I'll eliminate that possible outcome. If you can make a reasonable first guess and get the signposts right, that's a powerful tool for figuring out where you're going. This same technique can be applied to interest rate policy and other aspects of economics. Whether it's SDRs or gold or multiple reserve currencies or some kind of collapse, powerful analytic tools are available to help in predictive analytics and problem solving.

When there is a collapse, however, you'll see a policy response consisting of draconian executive orders and account freezes. These freezes will not be limited to bank accounts, but will include mutual funds, exchange-traded funds, and other popular investment products.

After a collapse you might end up on a gold standard or a modified gold-backed SDR. Gold or SDRs are the two most likely outcomes, and perhaps a gold-backed SDR is the best of all possible worlds. There are two ways to get there: pretty, and ugly.

The pretty way of getting to a gold-backed SDR is to do so rationally and think hard about the problem. You set up working groups, committees, and technical study groups to achieve consensus, and you work country by country to get the policy changes needed to bring it about. This is a version of what the Eurozone is going through today with regard to structural reform in Greece and the Eurozone generally.

The ugly way is to just ignore it, let it collapse, and then get there by fiat, or by executive orders. That's a much messier, more costly way, and you might end up in the same place. If people lose confidence in the existing paper currency system, you're going to have to restore confidence. It's either going to be a new currency, which is the SDR, or the oldest form of money, which is gold.

Many investors and savers will lose money when the banks are closed, which is a good reason to acquire gold now. Gold in physical form outside the banking system is immune to bail-ins. I would not leave all my money in the bank or in trading positions such as stocks and bonds. You need some money in the bank as working capital. Still, I would have some in physical gold outside the banking system. Banking accounts are going to be "bailed in" and frozen during the next crisis.

Bail-in Possibilities

"Bail-in" is jargon for a situation where bank depositors do not get all their money back in the event of a bank failure. Maybe a portion of a bank deposit is insured and a small account gets repaid. Still, larger accounts in excess of the insurance would be converted into bank equity or wiped out entirely. It's the same with debt holders and bondholders of the bank. A bank is insolvent when it has no capital, and its liabilities in the form of deposits and notes exceed its assets. In that situation, there's not enough money to go around and the bank creditors and depositors take a haircut.

Those parties have their notes and deposits forcibly converted into equity with the hope that at some point in the future, the bank regains its health and the equity has future value. It's not what the creditors, depositors, and bondholders expected, but it's better than nothing. That's a bail-in; you the depositor, you the creditor, have been bailed in to the position of being an equity holder against your will.

A lot of people are surprised to learn this, yet bail-in has been the law in the United States since 1934. Prior to 1934, there was no insurance for depositors, and if a bank failed, you might have lost *all* your money.

The Federal Deposit Insurance Corporation (FDIC) began in 1934 and continues today. There has always been a limit on the insured amounts. The limit has increased over the years and currently is $250,000. This is a significant amount that certainly covers small depositors. Still, large depositors, which could include wealthy individuals, retirees with a lot of savings, corporate accounts, business accounts, or institutions, might well hold a lot more than that on account at the bank. If you sell a house for $1 million, you might get a $1 million check deposited in your bank account on the day it sells. Although that bank deposit might not stay in the bank for long, you're exposed for the time being. It has always been the case that amounts in excess of the insurance are at risk even though most people may not realize that or may take bank safety for granted.

The bank failure wave that happened in 2008 and the years following would have been much worse without government intervention. Will there be a bail-in when the next panic takes place? In the United States we have multiple regulators with a voice in any such decision: the FDIC, the Federal Reserve, the Treasury Department, and the Office of the Comptroller of the Currency (OCC). Depositors have been repeatedly warned about the possibility of bail-ins in the event of a future collapse. Either a bank collapse could cause a market collapse or the opposite—a market collapse could cause a bank to collapse if asset values are going down. That's how contagion works. It's not always a straight line.

We haven't had this much instability in the banking system since the 1930s. During the 1980s, the United States had a lot of savings and loan associations that were closed, but by and large, the depositors were kept whole, and few lost money. Most of those losses fell on bondholders and equity holders. The 1980s crisis was not so bad that the FDIC had to invade the savings of the depositors. Prior to the FDIC, depositor losses were routine. There were bank panics throughout U.S.

history. The point is that bail-ins and depositor losses would not be anything new in U.S. banking history; they just have not happened lately.

What is new in U.S. banking history is depositor and regulator awareness, probably because of the financial crises in Cyprus in 2013 and Greece in 2015. After Cyprus, a number of regulators, including those in Europe and the United States, said that bail-ins are now a template that will be used in future panics. Note that this template approach was ratified by the G20 and IMF at the G20 Leaders' Summit in Brisbane, Australia, in 2014. When the next wave of bank panics hits and the bail-ins occur on a mass basis, bank depositors will not be able to say they didn't see it coming. The prudent ones would have removed their deposits and bought gold in advance. That portion of their wealth will be preserved.

Confiscation and Windfall Profits Tax

With a major crisis in the monetary system, the U.S. government has the means to resort to gold confiscation or a windfall tax on gold profits measured in dollars, or both. If you're not a U.S. citizen, U.S. jurisdiction is limited, although the government could confiscate whatever is in storage on U.S. soil or in U.S. banks.

There are approximately six thousand tons of gold at the

Federal Reserve Bank of New York on Liberty Street in lower Manhattan. Little of that gold belongs to the United States; it belongs to foreign countries and the IMF, yet it could easily be confiscated by the U.S. Treasury to deal with an emergency economic situation. There are approximately another three thousand tons near JFK Airport, and there are a few other large storage locations including the HSBC vault at Thirty-ninth Street and Fifth Avenue in New York City.

It's possible that the United States could actually confiscate all that gold, convert it to U.S. government ownership, give the former holders a receipt, and tell them they can earn their gold back based on future compliance with new rules under a new U.S.-led international monetary system.

While it would be easy to confiscate foreign gold on deposit at the Federal Reserve and in other large vaults, it would be more difficult to confiscate gold held by individuals. In 1933, when FDR confiscated the gold of the American people, there was enormous trust in government, fear about the economy, and a belief that the president must know best. Many Americans felt that if the president told them to hand over their gold, they should just hand it over. Today those preconditions have changed. There is a lack of trust in government, and a view that politicians do *not* know best. There could be a lot of civil disobedience when it comes to handing over private gold held in physical form, so the government probably would not even try.

What the U.S. government may do, however, is enact a

windfall profits tax enforced through banks and dealers with mandatory reporting requirements. The government could require gold dealers to file transaction reports, cash reports, and 1099 forms, and create other information sources including federal licensing requirements. Using that information, the government could impose a 90 percent windfall profits tax on actual sales or even deemed paper profits (which would force many to sell their gold in order to pay the taxes).

Astute gold investors will be able to see this coming. A windfall profits tax on gold cannot easily be imposed by executive order because of congressional control over taxation. Such a tax requires legislation, and the legislative process is slow. Gold holders would know in advance and have time to prepare. It might not happen at all, because even a few members of the Senate would usually be sufficient to stop such legislation in its tracks.

A more serious threat than gold confiscation is a freeze on 401(k)s and mutual funds. There is a danger that the U.S. government could confiscate everyone's 401(k) in exchange for a government-backed annuity instead. Confiscation would simply convert the entire defined-contribution retirement plan system into an extension of social security. In a crisis, extreme measures cannot be ruled out, including confiscation, bail-ins, asset freezes, special taxes, windfall profits taxes, or swaps of stocks for annuities. In a crisis worse than 2008, every confiscatory technique you can think of is on the table.

When conditions become dire, world governments will do "whatever it takes." Too many analysts wear their free-market hats when they should be putting on their government hats and thinking like desperate bureaucrats. In the minds of government officials, the continuity of government power comes first and individual wealth takes a backseat.

Analysts take what they learned in business school or economics class and analyze politics in terms of rational markets and the efficient-market hypothesis. That's not how governments behave. Governments don't go down without a fight. If an economy is in distress, riots over money will be breaking out, people will be demanding their money back, and social unrest will go from bad to worse. A neofascist response cannot be ruled out.

War on Cash

In addition to the currency wars, and financial wars, there is also a war on cash in the United States. This assault on the right to freely use cash is another reason to consider an allocation to physical gold in times of economic uncertainty.

There are many good, legitimate reasons to hold cash. You might have a cash business. You might want to have cash for emergencies. If you live where I do, on the East Coast, we're vulnerable to hurricanes and nor'easters that can knock out power for days or weeks (as with Hurricane Sandy). When the power goes out, the ATMs and credit card readers don't work. It's good to have some cash around for occasions like that.

Nevertheless, we are witnessing an accelerated move toward digital currencies, or the so-called cashless society. People say, "So what? It just seems really convenient to go digital." I agree. I use credit and debit cards, PayPal and Apple Pay just like tens of millions of Americans. Yet the digital trend has some important implications.

An all-digital system sets up the economy for negative interest rates. The government could try to force people to spend money by confiscating part of whatever is left in a bank account under the guise of a "negative" interest rate. Instead of the bank paying you interest, they take it out of your account. Cash is an easy way to defeat negative interest rates. Anyone holding cash would have the same amount at the end of every time period and would not be subject to negative rates. Eliminating cash and forcing everyone into an all-digital system is the first step toward negative rates. Prominent economists Larry Summers and Kenneth Rogoff have advocated such steps.

The war on cash was launched ostensibly to pursue drug dealers and terrorists. The government authorities are always going to say, "We're not really against everyday citizens. We're just trying to get these bad drug dealers and these bad terrorists, tax evaders, and others. That's why we don't allow people to have cash." The problem is that law-abiding citizens are presumed to be drug dealers, tax evaders, or terrorists as soon as they express a preference for cash.

The war on cash is more than just a prelude to negative interest rates. Eliminating cash also makes it easier to force bail-ins, confiscations, and account freezes. To lock down depositors' money, it is helpful to herd everybody into one of a small number of megabanks (Citi, Wells Fargo, Chase, Bank of America, and a few others) that take orders from the U.S. government. Then the stage is set.

The war on cash is reminiscent of what happened to gold in the early part of the twentieth century, between 1900 and 1914. In the United States in, say, 1901, when you made a purchase, you might reach into your pocket and pull out a gold twenty-dollar coin or a silver dollar. I remember when I was a kid that a quarter or dime was still solid silver. It was only in the 1960s that the government debased silver coins by adding copper, zinc, and other alloys.

How did the government get people to give up their gold coins? Banks slowly took the coins out of circulation (the way cash is going out of circulation today), melted them down, and recast them into four-hundred-ounce bars. Nobody is going to walk around with a four-hundred-ounce bar in her pocket. Then they said to people, in effect, "Okay. You can own gold, but it's not going to be in the form of coins anymore. It's going to be in the form of these bars. By the way, these bars are very expensive." That means you needed a lot of money to have even one bar and you weren't going to take it anywhere. You were going to leave it in a bank vault.

It was a gradual process, and people didn't seem to notice the substitution of paper money for gold coins because the paper seemed more convenient (the way digital seems more convenient today). Banks created these four-hundred-ounce gold bars and got rid of gold coins. By the time gold was made illegal in 1933, there was not much gold still in circulation. It was relatively easy to confiscate the gold bars in the bank vaults using an executive order.

The same process is going on today. People are accepting digital substitutes for paper currency because it's more convenient. Only later, when paper money is completely out of circulation, will the government start to confiscate the digital wealth. There will be no recourse to cash at that point. People won't notice until it's too late.

The entire process of shifting from gold coins, to gold-backed paper, to fiat paper, to digital money has taken about one hundred years. Each step in the process makes it that much easier for the government to confiscate your wealth.

We've now come full circle. I described the war on gold in the first half of the twentieth century. Now in the twenty-first century, we're seeing a war on cash. Ironically, the solution to the war on cash is to *go back to gold*, because gold is now legal again. From 1933 to 1975, gold ownership was illegal in America (and still is in many countries abroad). Yet gold is now a legal form of money ownership. You can buy large four-hundred-ounce bars if you want. You can buy one-kilo bars, which are a lot more convenient than four-hundred-ounce bars, and you can also buy gold coins. The U.S. Mint sells the American Gold Eagle and American Buffalo one-ounce gold coins. Both contain one ounce of pure gold, but the Eagle also adds an alloy for durability.

It probably is prudent to get cash in addition to gold. Still, that has become more difficult. You can start by going to your bank and asking for $5,000 in cash. It's not illegal, but you will be required to show ID, sign some forms, answer questions, and have reports filed with the government—probably a Form SAR, for Suspicious Activity Report. Amounts of $10,000 or more require a Currency Transaction Report, Form CTR. There's an automated reporting network behind SARs and CTRs, and red flags filter up to the U.S. Financial Crimes Enforcement Network (FinCEN), located in northern Virginia near the U.S. intelligence agencies. If you ask for cash, you may not be a drug dealer, yet you will be treated like one by your bank.

It's probably too late to get much cash. The war on cash is mostly over and the government won. Still, it's not too late to get gold, which maintains its worth as a physical store of wealth and is not affected by the digitization of other forms of money.

Retracement

Another reason to credit gold's resilience, even after significant price declines from 2011 to 2016, is the classic commodities trading pattern called retracement.

In the winter of 2015, I spent a couple of days in the Dominican Republic with Jim Rogers, the celebrated investor, commodities trader, and cofounder of the Quantum Fund along with his then partner George Soros.

Few investors have seen as many bull and bear cycles in as many markets as Jim has. When we met, gold was about $1,200 per ounce; it later fell to the $1,050 per ounce level. Jim told me he was holding on to the gold he already owned, but was not accumulating more at the $1,200 level. He was waiting for what he described as a "50 percent retracement" and said that would be the signal to consider buying more. This technical trading approach did not alter our shared fundamental view that gold was ultimately headed much higher—perhaps to $10,000 per ounce or more. The idea of retracement had to do with short-term trading opportunities and entry points for new investment, rather than the fundamental case for gold.

For example, if gold formed a base of $200 per ounce, which it did in the late 1990s, and then rose to $1,900 per ounce, which it did by August 2011, a 50 percent retracement would take gold back down to $1,050 per ounce, which is the midpoint between $200 and $1,900. Jim told me he had never seen a commodity rise from a low base to a super-spike without a 50 percent retracement along the way. The bottom line for Jim Rogers was to begin adding to his gold portfolio around the $1,050 level.

This kind of price action or volatility is not unusual, yet the long-term forecast has not changed because the economic fundamentals and the monetary math are the same. The endgame for gold in a world that loses confidence in paper money is at least $10,000 per ounce, possibly much higher, because a higher price is needed to restore confidence in a panic without causing deflation. If you believe panic is unlikely in the future or that confidence in paper money will be maintained indefinitely, then perhaps gold is not your preferred type of money. History shows that panic and lost confidence are just a matter of time. In those circumstances, gold is the safest store of value.

Despite investor concern over retracement, which is not surprising, gold still has an excellent long-term track record measured in dollars, up 450 percent since 1999, and up 3,000 percent since 1971 as of this writing.

One danger for small investors is that as we get closer to the point where gold rises sharply, it may become increasingly difficult to obtain physical gold. I don't doubt that the central banks, sovereign wealth funds, and major hedge funds are going to be able to get some physical gold at a price. Yet one can see a time when the mints are going to stop shipping, small dealers are going to be back-ordered, and you're not going to be able to get gold in the quantities you want regardless of price. That's another reason it's a good idea to acquire a gold allocation now, so you're not left out in the cold when the price begins to gap upward in an uncontrolled and disorderly way.

Conclusion

As a twenty-first-century investor, I don't want all my wealth in digital form. I want part of my wealth in tangible form, such as gold. You can't hack gold, you can't digitally delete or erase gold, and you can't infect it with a computer virus, because it's physical.

Given the turbulence afflicting the international monetary system today through currency wars, cyberfinancial wars, and the war on cash, my forecast for gold remains that it is headed for a much higher dollar price in the not-distant future. The economic circumstances and conditions that support this analysis have not changed. Gold's resiliency in a time of turmoil has been proven time and again.

Chapter 6

HOW TO ACQUIRE GOLD

The Gold Market

The gold market is unusual compared with other markets in stocks, bonds, or commodities. On the one hand, it's liquid—an investor can easily count her gold allocation among her liquid assets. Liquidity means one can buy and sell gold with relative ease and minimal market impact. Your transaction will not stall the market and you will not have a hard time finding a buyer or seller to transact with you.

On the other hand, the gold market is a thin one. To be thinly traded means that the volume of gold trading relative to the total volume of gold is fairly small. What makes this unusual is that thin markets are usually not that liquid. In the case of gold, one finds a market that's not deep by volume. Still, it is liquid in terms of parties willing to transact. This is an unusual combination.

I've never seen a situation where there wasn't a bid for gold if you wanted to sell or where you couldn't buy gold if you wanted. Yet there's not that much physical gold that's actually traded relative to the total stock of gold in the world. Thin trading in gold is attributable to the fact that most gold holders, whether you're talking about central banks or brides in India, are long-term holders who are not looking to make a quick trade like so many stock and currency investors.

This means that the liquidity we see today could easily dry up in a buying panic. Millions would suddenly clamor to buy gold, but the long-term holders would refuse to sell even as the price soared upward. Normally high prices cause sellers to step forward to reestablish equilibrium. That's the basic supply and demand function we all learned early in our economics courses. But where "higher prices" actually represented collapsing confidence in currencies, no amount of paper money would be enough to bring the gold out of strong hands. Higher prices might make them *less* willing, rather than more willing to sell, because they would see the endgame of total collapse. At that point, only massive sales of government gold or a new fixed price for gold would be sufficient to stop "panic buying."

What could cause such panic buying in the real world?

We could have a credit crisis and a run on the banks in China. We've already seen signs of this in China's stock market collapse beginning in mid-2015. China still has a mostly closed capital account that prevents small Chinese investors from investing abroad. The everyday Chinese investor has already lost money on stocks and real estate, and China's banks don't offer significant interest on their savings. So what else could a Chinese investor do? She could buy gold.

I met recently in Hong Kong with the head commodities trader of one of the largest banks and gold dealers in the world. He foresees a demand shock coming from China that could segue into panic buying.

Another scenario is a failed contractual delivery by a gold dealer to a major client. When word of such a failure leaks out (which it always does), there could be a loss of confidence in the paper gold nexus and a mad scramble to convert paper gold contracts to physical gold by demanding delivery from futures warehouses and bank vaults. This would cause other delivery failures because there's just not enough physical gold to go around. Custodians and exchanges would resort to force majeure clauses in their contracts to cancel their delivery obligations and settle up in cash, yet that would just make the panic worse as investors realize they will never see their contractual gold and turn to other venues to buy physical gold as a substitute. That would quickly spiral out of control into a broader buying panic.

On a short-term technical basis I expect the gold price to go up gradually for a time, then that rate will accelerate to a hyperbolic phase, a super-spike—what we've been calling a buying panic. The problem is that a lot of people will want to jump on the bandwagon, but at that point, you might not be able to get the gold. There may come a time when the physical supply is so small that you can't buy it even at a much higher price.

As a result, investors ask, "How long do I have to buy gold before it runs out?"

Questions like that answer themselves. If you're asking how much time you have, the answer is: *what are you waiting for?* You should acquire physical gold now and put your mind at ease. Don't try to time the buying panic; by the time it's visible it will already be too late, and the small investor will not be able to get physical gold. It won't be a question of price. You're not going to be able to find it. The prudent course is to buy gold now, have it in a safe place, and when the gold buying panic comes, you'll be fine.

How to Acquire Gold

Storage

Decisions about your physical gold storage depend a lot on the quantity you hold. Are you talking about five one-ounce gold coins or $100 million in one-kilo gold bars? When you get into those large amounts, you probably need third-party custody unless you want to go to the expense of a home vault and multiple security perimeters.

My advice for third-party custody would be to go with private gold storage companies, not bank storage. The reason is that banks are heavily regulated, and gold held there can easily be confiscated by the state. I expect asset freezes and confiscations in a financial panic, so if you're in nonbank storage, you have a better chance of surviving that.

There are some highly reputable firms. Make sure you get references from other customers and check the certificate of insurance for the vault operator. Make sure the insurance limits are high enough and the insurance underwriter is a solid company. Consider how long the vault has been in business.

Any reputable vault operator will allow you to visit the facilities and inspect their security measures and procedures. For example, the internal loading dock should be positioned at right angles to the external doors so a vehicle cannot be used as a battering ram to break through both sets of doors.

The advice I give on picking a private vault is the same advice I give on almost any type of business decision. Look for people with a good reputation, look for people with long track records, and look for people with references, with insurance and bonding. Don't go with the fly-by-night vault that just set up shop. They might be okay, but how do you really know? Use only established custodians. Brink's, Loomis, and Dunbar are all established names in the United States. Still, there are many other smaller firms with good reputations and facilities. Neptune Global, Worldwide Precious Metals, and Anglo Far-East offer purchase and storage options where you own the gold directly, but they handle the details. (For those who want precious metals returns with less volatility than gold, Neptune Global offers a composite investment consisting of gold, silver, platinum, and palladium in specified percentages).

Another way to own physical gold without having to do custody yourself is to invest in a gold fund that offers secure logistics and immediate redemption in cash or gold. Technically you own units in a fund, not the gold itself. But these funds are not like ETFs, futures, or LBMA unallocated contracts. The kind of fund I recommend owns only gold, fully allocated, and will deliver it to you the next day in redemption of your units on request. Both the Physical Gold Fund and Sprott are recommended for this purpose.

As far as jurisdictions are concerned, there are trade-offs. Many investors debate the relative merits of Switzerland, Singapore, Dubai, and a few other favored jurisdictions. Most reject the United States because it is viewed as a country most likely to confiscate gold (it has done so before).

The problem is that the time you most want your gold will be the time when social conditions are unraveling most rapidly. There may be riots and power grid outages. If you live in the United States, how will you get to Switzerland to pick up your gold? My wealthy friends say they will use their private jets, yet that overlooks the fact that you cannot fuel a jet when the power grid is out and the gas pumps don't work. Switzerland may be secure in the short run (it is) yet may not be the most practical jurisdiction in the long run if transportation networks break down along with civil society.

Disaster planning is complicated. You can't look at just one or two bad outcomes; you have to look at all of them to see where you stand.

A prudent approach for large holders of gold is to put some on deposit with nonbank vaults in Switzerland (to protect against U.S. confiscation) and to keep some closer to home (to provide backup if Switzerland becomes unreachable). If your gold holdings are more modest, a nonbank depository near your home is the best option.

The State of Texas is building a state bullion depository outside the banking system. Texas will rely on state sovereignty and the Tenth Amendment of the U.S. Constitution to resist any efforts by the federal government to confiscate gold. That's an option definitely worth looking into. It's only a one- or two-day drive by car from most of the continental United States. In an extreme situation, you should be able to drive down to Texas, pick up your gold, and drive home before the highways are closed. If the highways are clogged, use a motorcycle.

Leaving aside logistical disaster planning, my favorite jurisdiction in the world for gold storage is Switzerland. U.S. citizens shouldn't even think about opening bank accounts in Switzerland. Even if you are entirely legal and fulfill the reporting obligations, it's still an invitation to enormous government scrutiny and suspicion. Besides, the Swiss bankers really don't want your business these days for the same reason. But physical gold storage accounts with nonbank custodians are not bank accounts, and they don't produce income, so there's no income tax to consider. Those storage accounts should be available to U.S. citizens.

U.S. citizens might want storage closer to home, however. It depends on how much gold you have. If you have twenty one-ounce gold coins—that's about $25,000 worth of gold—you don't necessarily need it to be in a Swiss vault. If you're looking at larger quantities such as four-hundred-ounce bars or kilo bars, I would certainly consider Switzerland. It has a good rule of law, political stability, neutrality, fine infrastructure, and a well-trained military, and has not been successfully invaded for more than five hundred years.

Singapore has a lot of the attractions of Switzerland in terms of being a good rule-of-law jurisdiction and politically stable, yet it's in a worse neighborhood. It's close to a lot of political instability in Thailand and Malaysia and uncomfortably close to Communist China. Australia is another good option, but of course, it's not close to anywhere. (Australia works fine if you live in Australia.)

In 2013, I visited the vaults in Switzerland near Zurich where the Physical Gold Fund actually stores its gold. I traveled with fund representatives, and we were accompanied by two partners from the Fund's auditors, Ernst & Young. We were there to inspect and audit the fund's gold held for investors.

The vault operator brought the gold out on pallets with a forklift and placed it on a pedestal for our inspection. Vault personnel broke the seals securing the wooden crate that held the gold and removed the top. Inside we saw the gold bars with serial numbers, dates, refinery stamps, assay stamps, and numbers indicating weight and purity. The auditors reconciled their printed ledgers to the actual bars and confirmed that all were present and accounted for. That was a fascinating process to watch. It is unfortunate that the U.S. government is not as transparent about its gold holdings as this private fund.

There are vault features that seem right out of a spy novel. For example, there are external garage doors that open to allow armored cars into the vault. You might think to yourself, why couldn't I just take a battering ram and smash down that door and break into the vault? Well, on the other side of that door is another door. They lower the first door behind you and open the second door in front of you. You pull into a second loading bay with cement and steel-reinforced barriers straight ahead, so if you tried the battering ram technique, you wouldn't get far. They off-load the gold from the truck at a ninety-degree angle, so any vehicle trying to ram its way through would quickly hit a dead end.

Even at that point, you're only beginning to get into the vault. There are special portals where the gold is deposited on one side, a steel door closes, and someone takes it out the other side. There are also security cameras, motion detectors, and concertina wire—a form of barbed wire with blades—all over the facility, not to mention Kevlar, bulletproof glass, armed guards, and multiple security perimeters. It's difficult to imagine its being more secure.

We met with vault officials and had revealing discussions. They're seeing a steady inflow from bank storage to private storage. In theory, bank storage is just as good in the sense that the vaults are just as secure, yet that's not the point. The point is that the banks are heavily regulated by world governments, and gold stored there might be subject to seizure by these governments—or at least it will be easier to seize when the time comes.

Banks may fail and your claim to any gold held there might get tied up in a court proceeding. Your gold also might be "bailed in" as it was in Cyprus or your gold might be considered an unsecured asset and used to recapitalize the bank. You might even get worthless bank shares instead of your original gold. In private storage, you don't have those issues. Because of these concerns, the private vault operators can't build new vaults fast enough.

We went down to one of the largest refineries in Switzerland and heard the same story. We were able to view the operation from the inside as well as have constructive and quite lengthy conversations with senior officials. Their refining process today is heavily automated, and they're working at maximum capacity around the clock. Despite these efforts, they still can't meet all the demand.

With refineries producing massive quantities of pure gold, where are they getting the gold to refine? There are three main sources. One is semirefined gold, so-called doré, which is about

80 percent pure gold coming from the miners; there is so-called scrap, which is just jewelry, necklaces, bracelets, watches, and other gold items coming from a variety of sources; and there are gold bars that are being converted into smaller bars. These four-hundred-ounce London Bullion Market Association good delivery bars are being turned into one-kilo bars to Chinese specifications.

The refineries are doing a lot of their business by melting down old four-hundred-ounce bars and turning them into new one-kilo bars. The refiners also improve the gold's purity from, say, 99.50 percent pure to 99.99 percent pure. This purity is known in the trade as "four nines" and is the gold standard refining required by the Chinese.

In effect, China is turning its back on the London Bullion Market Association and redefining "good delivery" in the world market. It's doing it with the Shanghai Gold Exchange, gold contracts on the Shanghai Futures Exchange, its own refineries, and its own specifications. Shanghai is becoming the center of the world gold market, leaving London behind.

Gold Mining Stocks

We have focused on physical gold and its paper derivatives such as ETFs and futures contracts. But gold is represented in the stock market too, in the form of gold mining stocks.

I've researched, written, and spoken a lot about gold, yet

my involvement is almost always related to the physical metal, the derivatives, and the use of gold as a monetary asset. I do not hold myself out as an expert in mining stocks; I'm not a stock picker. I'm not really an equity analyst in the traditional mold; I'm a global macro analyst. I think about gold in the monetary context. Still, capital markets are so interconnected today that the macro affects the micro more than ever. That's the basis for offering views on mining stocks.

Gold mining stocks follow gold to a great extent, but they're more volatile. Traditionally, gold mining stocks have been described as a leveraged bet on the physical metal. There are technical reasons for this, having to do with the difference between fixed costs and variable costs, but basically, when gold goes up, mining stocks go up even more. When gold goes down, mining stocks may underperform and fall faster than the metal itself. Gold mining stocks are like gold on steroids.

Gold is volatile enough as it is. Most investors won't have the appetite to add implied leverage to gold by going into the gold mining sector. If you want to leverage your gold position, you can get plenty of leverage using COMEX gold futures or buying an ETF on margin. That's one way to think about what it means to buy gold mining stocks.

Of course, what sets gold mining stocks apart as an asset class is that they're completely idiosyncratic. In other words, miners are never generic like futures or indices. Mining companies will each have unique characteristics related to the ore quality, the competence of management, corporate finance, and other one-of-a-kind factors.

One problem with investing in gold miners is that many investors do not treat the mining companies as idiosyncratic; they treat them generically. They say, "Oh, well, I invest in *miners*," as if they're all the same. They're not.

Some gold miners are well-run companies, established and solidly capitalized. They've been around for a long time, and one expects they'll continue to be around for a while. Other entities are highly speculative. Some are frauds.

As an investor, how do you sort out the frauds from the well-run companies? You can do it. Still, that's a lot of work. You probably have to meet with management, travel to the mines, read the financials (including the footnotes), go to investors' meetings, and listen in on management calls. Basically, you have to be an equity analyst, which is what I'm trained to do, but I don't usually do that kind of analysis because it's not my area of expertise.

There are experts doing solid research on gold miners, like John Hathaway and his colleagues at the Tocqueville Gold Fund, and Doug Casey and the folks at Casey Research.

I prefer the physical metal itself. If you want to be an investor in the gold-mining sector, I recommend sticking to the large-cap and selective mid-cap gold miners. Here's why: microcap or start-up miners are mostly in bad shape. We all know how that sector has performed; the stocks have been beaten down. Many of these stocks are down 95 percent from their highs. People say if they're beaten down enough, isn't that a good time to buy? Well, there's a better time to buy, which is *after* they go bankrupt. Master limited partnerships, roll-ups, and bigger firms looking for a bargain will cherry-pick the best assets from among those in distress.

The banks and big miners aren't out to do the small miners any favors. Investors tend to think you can "buy low" and get in on the upswing. Sometimes you can. Still, the savviest investors—larger mining companies and predators like Goldman Sachs—would prefer to see small miners driven into bankruptcy and their assets made available in a bankruptcy sale. As a prepetition equity holder in a leveraged miner filing for bankruptcy, you can expect to receive nothing.

Gold's Lack of Correlation with the Stock Market

A lot of investors think of stocks and gold as a trade-off. Their view is that when times are good—meaning strong economic growth, low unemployment, and price stability—they prefer stocks to gold. Yet when conditions are rocky, inflation is getting out of control, or the economy is highly uncertain, they prefer to dump stocks and have a safe haven like gold.

Some investors have even suggested, in effect, that I call

them the day before stocks collapse so they can sell their stocks and buy gold just in the nick of time.

Of course, it doesn't work that way; I'm not going to know the day. I do have insights into the magnitude of the coming collapse, and the consequences. The collapse is likely to come sooner than later, perhaps in months or in a year or two. We're unlikely to make it five years without a severe financial panic. Regardless of the specific date of the collapse, the time to prepare for it is now. I won't know the date any sooner than you will, yet I'm doing my best to prepare.

Until the panic begins, there is no particular correlation between gold and stocks. At times stocks and gold go up together (that's likely in the early stages of inflation), and at times gold goes up when stocks go down (in panics or the late stages of inflation). At other times gold goes down and stocks go up (in a strong economy with positive real interest rates). Finally, gold and stocks can both go down (in deflation). In short, gold and stocks move in different directions in response to varying conditions with no long-run correlation between them.

One could see, for example, the economy getting stronger in nominal terms with inflation picking up. In that scenario, stocks and gold go up together because stock investors would see higher revenues due to inflation and gold investors would hedge with gold in the expectation that inflation will get worse.

On the other hand, if inflation gets out of control and the

Fed gets behind the curve, the inflation begins to destroy capital formation and most forms of wealth. This could lead to stagflation, where we don't have real growth but we do have inflation. This would be similar to 1975–79. In that world, we would see gold going up for fundamental reasons.

So I don't see a correlation between stocks and gold. There is a place for both in a carefully selected portfolio. For example, I recommend certain hedge funds and alternative investments that have stocks in their portfolios. These funds include long-short equity funds and global macro funds that are making directional bets in a few sectors.

There is also nothing wrong with stocks that have underlying hard assets. A good example is Warren Buffett's buying of railroad, oil, and natural gas assets. Warren Buffett is a classic stock investor, but he's buying companies that have hard assets such as energy, transportation, and land underneath them, and those stocks should do just fine in an inflationary environment. The right portfolio is one that has a good mix of assets, with gold being about 10 percent of the mix.

Gold in a Well-Balanced Portfolio

My intermediate-term forecast for the gold price has not changed despite volatility and retracement in the nominal dollar price. Gold will ultimately be making its way to the $10,000 per ounce range. This will happen because either central banks will succeed in causing inflation, or they will fail and finally turn to gold as the inflation *numéraire* of last resort (as FDR did in 1933). Either way, central banks will eventually get the inflation they need to make debt levels sustainable. This condition was described by former Fed governor Rick Mishkin as "fiscal dominance."

A new gold standard or at least gold-linked currency system may be needed in a world of inflation and fiscal dominance where gold is called upon by central banks either to cause inflation or to restore confidence after inflation has gotten out of control.

The implied price analysis is straightforward. It's the ratio of paper money to official physical gold in the world. The official gold hoards and official money supply figures are both known (with exceptions such as China's off-the-books hoard).

Some assumptions are needed. For example, which countries will be included in a new gold standard? What is the proper definition of money supply (for example, M0, M1, M2, et cetera)? What ratio of gold to money will be required to maintain confidence in a new gold-linked system (for example, 20 percent, 40 percent, et cetera)? Subject to gathering the data and making those assumptions, the math is straightforward. An indicative nondeflationary price of $10,000 comes from an assumption that the United States, the Eurozone, and China will all be included, and M1 with 40 percent gold backing will be the right measure of money and gold. Other assumptions will produce different results—some as high as $50,000 per ounce if one uses the M2 money supply with 100 percent gold backing. We're not there yet—it may even be several years away—but that's where I see it going. Still, it can be quite a tortuous path.

In the meantime, gold investors should follow a few simple rules. Gold is volatile when measured in dollars, so I recommend not using leverage in gold investing. When you use borrowed money, margin, or go into the futures or options market, you're using leverage that will amplify the underlying volatility. Gold is volatile enough as it is and doesn't need extra volatility on top.

Second, I've consistently recommended a modest allocation—10 percent of your investible assets for most investors, or 15 to 20 percent of your investible assets if you're somewhat more aggressive. I've never said, and I'm not going to say now, that you should sell your entire portfolio and buy gold. I don't believe in going 100 percent into one asset class.

By way of comparison, institutional allocations to gold around the world are only about 1.5 percent. Even if you took my conservative recommendation of 10 percent and you cut that in half to 5 percent, that's still more than three times what institutions actually have in gold.

The 10 percent allocation is meant to apply to your investible assets—the liquid part of your portfolio. You should exclude your principal residence and any equity in your business from the investible asset pool. Perhaps you have a restaurant, dry cleaner, or pizza parlor, or you're a car dealer, doctor, or dentist. Whatever capital is tied up in how you make a living should not be included in the "investible asset" pool. The same rule applies to home equity. Whatever is left after the business and home are separated is your investable assets. I recommend putting 10 percent of that amount into gold.

If you have 10 percent of your portfolio in gold and it goes down 20 percent, you've lost only 2 percent on your portfolio. That's hardly a wipeout. Still, if it goes up 500 percent, which I expect, then you'll do quite well on that 10 percent allocation. That's a 50 percent gain on your portfolio from one investment. I recommend the 10 percent allocation because of the asymmetry in the potential upside versus the potential downside. With these simple rules as a guide—buy physical gold, avoid leverage, and keep your allocation to 10 percent—you're ready to weather the storm.

Another important piece of advice is to stay focused on the long-term and don't get distracted with day-to-day ups and downs in the dollar price of gold. We already know it's volatile. More important, the dollar itself is the asset that's under threat, not gold. The fact that the dollar price of gold went "up" or "down" on a given day will come to be seen as irrelevant once confidence in the dollar itself evaporates. No one will care about dollars at that point—they'll just want physical gold.

CONCLUSION

My models on systemic risk in capital markets point toward dire events, including the collapse of the international monetary system in the not-distant future. Still, we are not helpless. We don't have to be victims. We can see the collapse coming and take steps in advance to preserve wealth.

There are always actions for investors to take and particular asset classes, especially gold, that are robust to extreme events. Certain investment strategies can be used, first of all, to preserve wealth and then even prosper and gain wealth through the chaos. Investors like Warren Buffett and others have done so for decades, and the Chinese are doing it now with their gold acquisitions.

I view risk through the lens of complexity theory and estimate events using mathematical models based on inverse probability. The most important complexity theory metric is scale. What's the scale of a system? *Scale* is a technical word for the combination of the size of the system and its density function or degree of connectedness given its size.

The most extreme event that can happen in any critical-state complex system is an exponential function of scale. This means that if you double the systemic scale, you *more* than double the risk. The risk can go up by many times the increase in scale.

Since 2008, banks have vastly increased the scale of the financial system, and regulators have turned a blind eye. The 2008 "too big to fail" banks are now bigger, and their derivatives books are much bigger than they were in 2008. This is the equivalent of sending the Army Corps of Engineers out to make the San Andreas Fault bigger. We know that the San Andreas Fault is an earthquake risk and that extreme earthquakes can happen. We don't know the timing of the next big earthquake, yet no one thinks it's a good idea to make the fault line bigger. That is what we're doing in the financial sector—making the fault line bigger by allowing the banks to get bigger, allowing derivatives to get bigger, and allowing a greater concentration of financial assets in few hands.

Using these simple scale and density inputs in a complex system, it is easy to see that the next financial collapse will also be exponentially bigger than the panic of 2008, because the system within which it takes place has grown enormously. Regulators simply do not understand the statistical properties of risk in the financial systems they regulate. They're using flawed equilibrium models that involve normally distributed risk (for which there is no empirical support). The regulators' risk management toolkit is obsolete.

My advice to investors is simply to get gold—but not too much. If I'm wrong, you won't be hurt by a small allocation. If I'm right, your wealth will be preserved.

Why wouldn't you go all in? Why wouldn't you have 50 or 100 percent of your portfolio in gold? The answer is it's *never* a good idea to go all in on any asset class. Gold is volatile, so you'll want other assets in your portfolio such as cash to damp down the volatility. Also, there are other assets that will preserve wealth the way gold does. Fine art and land have many of the same wealth preservation properties as gold, but still add diversification to a portfolio.

If institutions that currently have about 1.5 percent in gold simply went to a 5 percent allocation, lower than my recommendation, there's not enough gold in the world at current prices to come anywhere near filling that demand. That's an example of how fragile the system is and how little it would take to send the gold price soaring.

You don't need the end of the world for a super-spike in the gold price to happen. You just need fairly small changes in the behavior and perception of potential gold buyers for the gold price to skyrocket. After that, the feedback loops kick in and the gold buying frenzy takes on a life of its own.

I'm not predicting tomorrow's gold price. It could be up or down. I'm looking several years down the road to form an estimate of how the next panic will play out. I know that when the crunch comes, the large players are going to get all the gold available. The institutions, the central banks, the hedge funds, and the customers with relationships with the refiners are the ones who are going to get all the gold. Small investors will find they can't get any.

Your local dealer will be sold out and back-ordered. The Mint will stop taking orders. Meanwhile, what's happening with the price? It's going up more than $100 an ounce per day, more than $1,000 per ounce per week. It's running away from you. You want it yet you can't buy it. That is what a buying panic looks like.

Panic is the likely outcome of forces in play today. My best advice is to get your gold now while you still can. Then just sit tight and don't worry about it. Allocate 10 percent of your liquid assets to gold, put it in a safe place, then sit back and watch the show. It won't be pretty to watch; yet your wealth will be intact.

BONUS CHAPTER

FIVE STOCKS THAT COULD SHINE IN *THE NEW CASE FOR GOLD*

BY DAN AMOSS, CFA

In *The New Case for Gold*, Jim outlines many actions you can take to prepare your portfolio for a collapse of the international monetary system. In a monetary collapse, demand for gold would soar; the market for physical gold would tighten; and it might become very difficult to acquire any gold.

Besides holding physical gold, what's another way to bolster your portfolio with a secure claim on physical gold? And where will investors turn if they can't buy any physical gold during a shortage?

One of the first places investors will turn is the gold sector of the stock market. Gold stocks are claims on gold royalties, gold streams or gold mines. The value of carefully selected gold stocks could, as Jim explained, rise at a faster rate than gold itself.

The potential for gold stocks to outrun gold is unusually good today. Here's why: Since 2011, gold stocks have suffered through a bear market that set records in terms of its ferocity and length. Most gold stocks are down anywhere from 60–90% from their 2011 peaks. Valuations for well-positioned gold stocks are cheap, and expectations are very low.

Many junior exploration-stage stocks and even some producers will disappear. But many will also thrive and have been investing in smart royalties, streams, acquisitions and mine developments. Such investments will create enormous future value for shareholders willing to wait patiently for the next gold bull market.

We'll list five specific gold stocks that Jim and I have worked together to identify. All five are poised to thrive in a stronger gold price environment while also having the financial resources to ride out a bear market that lasts for several more years.

But first, let's consider what might happen to gold stocks if the U.S. government targets the wealth of gold holders . . .

Like any other stock, gold stocks are claims on corporate assets. There is no legal precedent for gold stocks to be confiscated without due process or fair compensation. Congress would have to pass a new law specifically targeting the rising profits of gold stocks. A new law is very unlikely; gold mining is not a politically sensitive industry like the oil industry, which was subject to a "windfall profits tax" from 1980–88.

Beyond a handful of jewelry manufacturers, there is no real political coalition of gold buyers. So there will be no political consensus to stop higher gold prices from benefitting gold producers at the expense of gold buyers.

Besides, many gold buyers will feel better about—not harmed by—buying gold in a rising gold price environment. Gold is money, as Jim explained in Chapter 2; gold is not a typical consumer good subject to the conventional laws of supply and demand.

Demand for gold is more likely to rise than fall in a rising price environment, because the public will start viewing gold as a more broadly accepted store of wealth. Shareholders of gold companies will not be seen as profiting "at the expense of" the American public.

In a gold futures market breakdown scenario, in which there would be a cash settlement in lieu of physical gold delivery, gold stocks would likely move sharply higher. Investors would view gold in the ground owned by well-run mining companies or royalty companies as extremely valuable options on the future gold price. And such options wouldn't involve the same risks as a gold futures market that's been tainted by cash settlement.

With such a promising future for a gold sector that's been cheapened by a bear market, we've selected five specific stocks that will survive and thrive:

Franco Nevada (FNV: NYSE and TSX)

Franco Nevada is a royalty company. It's our favorite gold stock for buy-and-hold investors. A royalty company spends investors' money to finance gold miners' operations. In return, the royalty company gets a percentage, or "royalty," from the mine's revenue. The painful bear market in gold mining has been a great environment for royalty companies like FNV to deploy capital on highly favorable terms.

It can deploy its own capital, plus loan proceeds, into value-creating precious metals royalties or streams. Streams have become a popular way for cash-strapped miners to sell the precious metals that often get produced as byproducts at base metal mines.

In a typical streaming deal, Franco Nevada pays cash upfront for the right to buy all the gold or silver produced at a specific mine at very low fixed prices for the life of the mine. Streaming deals are effectively long-term call options on the gold price, with very long expiration dates.

Royal Gold Inc. (RGLD: NASDAQ)

Like Franco Nevada, Royal Gold is one of the top gold royalty companies in the world. Its management team is top-notch, its assets are attractive and diversified and it has the potential to soar in a rising gold price environment. As gold prices rise, Royal Gold's business model and earnings growth kick into high gear, creating wealth for shareholders at a rapid pace. RGLD stock was brutally punished in late 2015.

Shareholders needlessly panicked about the state of a few of Royal Gold's assets. We've reviewed these assets and we're confident that the decline in the stock price is excessive and that management will calm investors' fears in 2016. Royal Gold has sowed the seeds of future shareholder wealth during the gold bear market. It struck a large number of new streaming and royalty agreements. Those agreements will become increasingly profitable in a rising gold price environment.

Agnico Eagle Mines (AEM: NYSE)

Agnico Eagle Mines is a blue chip gold mining company. Management has guided the company through the gold bear market in a calm, intelligent fashion, with a focus on building shareholder value. AEM holds a great portfolio of early-stage gold projects, along with eight producing mines in Canada, Finland and Mexico that produced over 1.6 million ounces of gold in 2015.

Thanks to low costs, AEM generates free cash flow—even in a depressed gold price environment. Management is prudently reinvesting this cash flow into stakes in promising junior mining stocks, which sets the stage for production growth in a rising gold price environment.

Alamos Gold (AGI: NYSE)

Alamos Gold is a company poised to thrive in a gold market rebound. Its key producing mines are in Mexico and Canada. Last year, Alamos closed a merger with a similar-sized peer, AuRico Gold. The merger was shrewd.

It strengthened the combined company's balance sheet and boosted its efficiencies of scale. And it combined two complementary gold portfolios. Now its two flagship operations are AuRico's Young-Davidson mine in Ontario, Canada, and Alamos' Mulatos mine in Mexico.

Both mines generate plenty of cash flow and have long production lives. Management is investing in projects that will lower its production costs. Alamos has the balance sheet capacity and cash flow strength needed to fund its planned investments.

Over time, the merger should diminish the valuation discount investors assign to smaller gold producers. Bigger producers enjoy better access to capital and a broader menu of potential growth projects.

Premier Gold Mines (PG: TSX; PIRGF: OTCBB)

Premier Gold Mines is a small-cap gold miner with five projects. All projects are in safe mining jurisdictions like Nevada and Red Lake, Ontario. Premier's executives have laid the groundwork for a big surge in production and earnings in the next bullish gold cycle.

They've wisely limited its financial risk by bringing in joint venture partners for some of its projects. It's allowed Premier to maintain a strong balance sheet. Premier is transforming from an early-stage explorer to a gold producer; the South Arturo mine in Nevada, which is 40% owned by Premier, is commencing production in 2016.

How high could these five stocks rise in a stronger gold price environment? It's not a stretch to expect the first three stocks—Franco Nevada, Royal Gold and Agnico Eagle—which have lower risk, to double if gold rebounds to $1,500. And at $1,500 gold, Alamos and Premier, which are higher-risk stocks, could triple or quadruple.

If gold rises to $10,000 or above—the levels Jim estimated for a responsible return to a gold standard—the sky is the limit. These stocks could rise 10-, 20- or 30-fold.

Dedicating a small space in your portfolio for these five gold stocks can help maintain and even build your wealth in a global monetary crisis.

ACKNOWLEDGMENTS

This book began as a series of online audio interviews, which were then transcribed, edited, rewritten, and finally shaped into what you are reading. I am immensely grateful to the interview sponsor, the Physical Gold Fund, and its principals and associates—Alex Stanczyk, Simon Heapes, Nestor Castillero, and Philip Judge—without whose support this book would not have emerged.

An interview requires an interlocutor, and for my Physical Gold Fund interviews I could have done no better than to have Jon Ward as my intellectual sparring partner. Not all interviews are created equal. In my experience, the best questions elicit the best answers, and Jon's questions were carefully framed and well thought out. We would not have had the content for the interview series, let alone a book, without his skillful inquiries.

As always, I am indebted to my super-agent, Melissa Flashman, and my publisher, Adrian Zackheim, for supporting my writing efforts before I had written my first book, and ever since. They are a constant source of confidence and inspiration to keep writing.

Adrian's team at Portfolio/Penguin Random House is as much a part of this book as I am. My editor, Niki Papadopoulos, lived up to the higher calling of muse as she encouraged, cheered, and cajoled me past the endless rounds of editorial decisions and deadlines that make publishing about much more than writing. That said, she also helped me to be a better writer with her succinct and apt questions in just the right places. The others on the Portfolio team, Leah Trouwborst, Stefanie Rosenblum, Will Weisser, Tara Gilbride, and Kelsey Odorczyk, were also a source of support and greatly appreciated.

Freelance editors are invaluable because they focus exclusively on the content without the business distractions of working for a publisher. The book benefited greatly from two of the best. Zach Gajewski took the raw transcripts and organized them into a first draft. Will Rickards took the late drafts and turned those into a better book. Zach and Will were like relay runners passing the baton from the early to final stages. I cannot thank them enough for their attention to detail and focus on quality.

Some writers may work in a cloister, but I'm not one of them. Family and friends surround me, and life is so much better for it. The constant interactions, debates, and give-and-take enrich the creative process. The love of your family is the ultimate safety net for the writer on a high wire. Thank you, Ann, Scott, Dominique, Ali, Will, Abby, Thomas, Sam, and James (and the canine corps of Ollie and Reese). I love you all.

INDEX